Volunteer Tourism

Volunteer tourism is one of the major growth areas in contemporary tourism, where tourists for various reasons seek alternative goodwill experiences and activities. To meet this demand there has been a surge in volunteer programmes offered in a range of destinations organised by a variety of charities and tour operators, which is predicted to continue to grow in the future.

Volunteer Tourism provides an in-depth analysis of the complex issues associated with traditional and contemporary volunteer tourism. Reflecting the growth in this phenomenon, this book provides a cohesive collection of chapters written from a range of international scholars. The theoretically rich, practically applied and empirically grounded contributions are based on current and diverse research in the area. This groundbreaking volume explores topics which have not been addressed in the literature before, such as the impact on host communities, introducing new areas and ideas to the field. A diverse range of themes are identified and addressed, including volunteer tourism and sustainability to, uniquely, the examination of volunteer tourism stakeholders – volunteers themselves, the host-to-guest exchange, and the organisations and management of volunteers. These themes are examined in a range of international case studies, demonstrating the wide range of issues associated with volunteer tourism.

This volume is a timely addition offering an innovative approach to the area. *Volunteer Tourism* will be of interest to both undergraduate and postgraduate students and researchers interested in tourism, as well as non-academics, practitioners, NGOs and government officials at all levels.

Angela M. Benson is Divisional Leader for Tourism at the University of Brighton, Adjunct Associate Professor, University of Canberra, and Founding Chair of the ATLAS Volunteer Tourism Research Group.

Contemporary Geographies of Leisure, Tourism and Mobility

Series Editor: C. Michael Hall
*Professor at the Department of Management, College of Business & Economics,
University of Canterbury, Private Bag 4800, Christchurch, New Zealand*

The aim of this series is to explore and communicate the intersections and relationships between leisure, tourism and human mobility within the social sciences.

It will incorporate both traditional and new perspectives on leisure and tourism from contemporary geography, e.g. notions of identity, representation and culture, while also providing for perspectives from cognate areas such as anthropology, cultural studies, gastronomy and food studies, marketing, policy studies and political economy, regional and urban planning, and sociology, within the development of an integrated field of leisure and tourism studies.

Also, increasingly, tourism and leisure are regarded as steps in a continuum of human mobility. Inclusion of mobility in the series offers the prospect to examine the relationship between tourism and migration, the sojourner, educational travel, and second home and retirement travel phenomena.

The series comprises two strands:

Contemporary Geographies of Leisure, Tourism and Mobility aims to address the needs of students and academics, and the titles will be published in hardback and paperback. Titles include:

The Moralisation of Tourism
Sun, sand…and saving the world?
Jim Butcher

The Ethics of Tourism Development
Mick Smith and Rosaleen Duffy

Tourism in the Caribbean
Trends, development, prospects
Edited by David Timothy Duval

Qualitative Research in Tourism
Ontologies, epistemologies and
methodologies
*Edited by Jenny Phillimore and
Lisa Goodson*

The Media and the Tourist Imagination
Converging cultures
*Edited by David Crouch, Rhona Jackson
and Felix Thompson*

**Tourism and Global Environmental
Change**
Ecological, social, economic and political
interrelationships
*Edited by Stefan Gössling and
C. Michael Hall*

**Cultural Heritage of Tourism in the
Developing World**
Dallen J. Timothy and Gyan Nyaupane

**Understanding and Managing Tourism
Impacts**
Michael Hall and Alan Lew

Forthcoming:

**An Introduction to Visual Research
Methods in Tourism**
Tijana Rakic and Donna Chambers

Volunteer Tourism

Theoretical frameworks and practical applications

Edited by Angela M. Benson

Routledge
Taylor & Francis Group

LONDON AND NEW YORK

First published 2011
by Routledge
2 Park Square, Milton Park, Abingdon, Oxon OX14 4RN

Simultaneously published in the USA and Canada
by Routledge
270 Madison Avenue, New York, NY 10016

*Routledge is an imprint of the Taylor & Francis Group, an informa
business*

© 2011 Selection and editorial matter: Angela M. Benson; individual
chapters, the contributors

Typeset in Times New Roman by
GreenGate Publishing Services, Tonbridge, Kent

British Library Cataloguing in Publication Data
A catalogue record for this book is available from the British Library

Library of Congress Cataloguing in Publication Data
Volunteer tourism / edited by Angela M. Benson.
p. cm.
Includes bibliographical references and index.
1. Volunteer tourism. I. Benson, Angela M.
G156.5.V64V65 2010
361.3'7--dc22
2010026540

ISBN: 978-0-415-57664-2 (hbk)
ISBN: 978-0-203-85426-6 (ebk)

For my dad

Contents

Figures

Tables

Contributors

Zoë Alexander is a PhD student at Buckinghamshire New University, UK, under the guidance of Dr Ali Bakir and Dr Eugenia Wickens. Her research interest is in 'the self' (the tourist) and the impact that tourism has on the individual. Her academic background is psychology, a subject she has taught in adult education. More recently, she completed her MSc in Tourism Management and Development to support her business offering self-catering units in Cape Town and Scotland. Her commercial background is project management and she has worked for a large retailer in the UK. Her long-term personal interests lie in the development of self and hiking. Email: zaalexander@yahoo.co.uk or zoe.alexander@bucks.ac.uk

Professor David B. Arnold is Director of Research Initiatives and Dean of Graduate Students at the University of Brighton, UK. He has been involved in 30 years of research in the design of interactive computer graphics systems and their application in architecture, engineering, cartography, scientific visualisation, health and more recently cultural heritage and tourism. He is co-ordinator for 3D-CoForm, a large-scale integrating project under the EU Framework 7 programme. He was educated at the University of Cambridge and has an MA in Engineering and Computer Science and a PhD in Architecture. Email: D.Arnold@brighton.ac.uk

Dr Ali Bakir is Principal Lecturer at the School of Applied Management and Law, Faculty of Design, Media and Management, Buckinghamshire New University, UK. He lectures on Strategy and Marketing in Sport, Leisure, Tourism and Music. He also leads the school's postgraduate programme in Sports Management and the MBA (Sport). Ali's research interests lie in interpretive studies in Strategy and the Creative and Cultural Industries. Email: Ali.Bakir@bucks.ac.uk

Dr Angela M. Benson is Divisional Leader for Tourism and Travel and a Principal Lecturer in Tourism at the School of Service Management, University of Brighton where she has been since January 2004, having previously held the position of Senior Lecturer at the Southampton Solent University (1995–2004). Prior to her career as an academic, she worked for 13 years in leisure and recreation, managing a range of facilities and events and was responsible for working with sporting volunteers. Angela has published articles in the areas of Volunteer Tourism, Best Value, Sustainability and Research Methods. She is the Founding Chair of the Association for Tourism and Leisure Education (ATLAS) Volunteer Tourism Research Group. Current research is being undertaken with Australian colleagues into the legacy of volunteering in mega sporting events at the 2010 Olympic and Paralympic Winter Games. Email: amb16@brighton.ac.uk

Associate Professor Dr Jennifer Kim Lian Chan has a PhD in Tourism and Hospitality Management and is a tourism lecturer at the School of Business and Economics. She is also the Deputy Director for the Centre for Academic Advancement of Universiti Malaysia Sabah, Malaysia and a national panel auditor for the Tourism and Hospitality Program appointed by the Malaysian Qualifications Agency. Her research interests and current research areas include nature tourism management/ecotourism, tourist/guest behaviour (tourist experiences, satisfaction and service quality in tourism), service experience management, small and medium-sized accommodation management and tourism destination marketing and positioning. Email: jkimchan@ yahoo.co.uk; jenniferchan@ums.edu.my

Bilge Daldeniz is currently working as Research Associate at the Centre for Tourism in Islands and Coastal Areas (CENTICA) at the University of Kent, UK, after six years as researcher and consultant at a private sector international consultancy firm. Her research area is tourism in developing countries with a particular focus on Latin America and the Caribbean. She holds an MRes in Environment and Development from Lancaster University and is currently researching the impacts of climate change on tourism businesses and their response and adaptation strategies. Having worked herself as a volunteer on social and environmental projects abroad in the past, she also developed a research interest in volunteer and charity tourism. Email: B.Daldeniz@kent.ac.uk

Associate Professor Tracey J. Dickson works for the University of Canberra where she has been involved in researching alpine tourism, snowsports injuries and risk management in outdoor recreation. Her interest in volunteers and tourism draws upon her experience working in outdoor education and recreation where many volunteers help make the participants' experience memorable. Tracey's interest in volunteers and tourism has led to her conducting research on the legacy of volunteering in mega sporting events such as the 2010 Olympic and Paralympic Winter Games. Email: Tracey. Dickson@Canberra.edu.au

Liam Fee is a projects officer and PhD candidate in the Department of Development and Economic Studies, University of Bradford, UK. He is Chair of Trustees of Village-to-Village, a small, non-profit, UK organisation which provides volunteer-tourism placements through its partner, Village-to-Village, Tanzania. Email: L.Fee@bradford.ac.uk

Simone Grabowski is a Research Assistant in the School of Leisure, Sport and Tourism at the University of Technology, Sydney (UTS) where she graduated with a first class honours degree in Tourism Management. She is currently teaching and researching in the areas of sustainable tourism, volunteer tourism, protected area management and community development. Email: Simone.Grabowski@uts.edu.au

Dr Mark P. Hampton FRGS is Senior Lecturer in Tourism Management at the University of Kent, UK. He has extensive field experience in South-East Asia, particularly Indonesia and Malaysia, and has also worked in the Caribbean, South Atlantic and Europe. His funded research projects include backpacker impacts in South-East Asia (funded by the Ministry of Tourism, Malaysia); cross-border tourism (British Academy); and dive tourism impacts (British Council). He is a Fellow of the Royal Geographical Society and Visiting Professor at Universiti Teknologi Malaysia. Email: M.Hampton@kent.ac.uk

Joanne Ingram is a doctoral candidate in the School of Asian Languages and Studies at the University of Tasmania, Australia. Combining her interests in travel and Asia, her area of research is volunteer tourism and development in Asia, with a particular focus on whether a relationship exists between the two. Prior to her current academic pursuits, Jo spent several years employed within both the public and charity sectors. She has extensive experience within Human Resource Development, including five years as a Training Manager for a leading UK charity. Jo is a graduate of the Chartered Institute of Personnel and Development (UK) and has an Advanced Diploma in Training and Development. Email: Jo.Ingram@utas.edu.au

Dr Steven Jackson is Tourism Programme Manager at Southampton Solent University, UK. Originally a physical geographer, with a PhD in soil erosion modelling, he moved into tourism via ecotourism which provides the link to the natural environment. Subsequently his key interests are in issues surrounding the impacts of tourism and their alleviation through the adoption of sustainable and responsible tourism philosophies. Having undertaken voluntary work for a number of environmental groups, notably the National Trust (as a working holiday leader) and the Royal Society for the Protection of Birds, recent research has been focused on the area of volunteer tourism. Consultancy work has been in the area of countryside recreation use. Email: Steven.Jackson@solent.ac.uk

Dr Jamie Kaminski is a Research Fellow at the University of Brighton Business School (UK) where he specialises in the study of the socio-economic impact of heritage. He began his career as an archaeologist and has a PhD in archaeology from the University of Reading (1995). He has a long-standing research interest in all aspects of the management of heritage sites, and their social, economic and environmental impact. Other research interests include the impact of social enterprise and the application of new technology to heritage. He is head of heritage research at the Cultural Business research group at Brighton Business School. Email: J.Kaminski@brighton.ac.uk

Dr Anna Mdee is a Senior Lecturer in Development Studies at the University of Bradford. She has researched and published on community-driven development, sustainable livelihoods approaches, NGOs and Social Enterprise in Sub-Saharan Africa. She is also the co-founder of Village-to-Village, an NGO working in Tanzania on sustainable agriculture, supporting people living with HIV/AIDS and community-based tourism. She has been organising volunteer travel experiences since 2005 in Tanzania. Email: A.L.Mdee@ bradford.ac.uk

Dr David Mittelberg is Senior Lecturer and former Head of the Department of Sociology at Oranim, The Academic College of Education, Tivon, Israel. Dr Mittelberg also serves as a Senior Research Fellow at the Institute for Kibbutz Research at Haifa University, Israel, where he formerly served as its Director. Dr Mittelberg has published articles on ethnicity, migration, gender, tourism, kibbutz education and the sociology of American Jewry. Email: davidm@oranim.ac.il

Professor Michal Palgi works at the department of Sociology and Anthropology and is Chair of the MA Program for Organizational Development and Consultancy at the Emek Yezreel College, Israel as well as a senior researcher at the Kibbutz Research Institute, The University of Haifa. She is president of the International Communal Studies Association, co-editor of the *Journal of Rural Cooperation* and former president of the International Sociological Association Research Committee of Participation, Organizational Democracy and Self-Management. Her main areas of interest and publications are: kibbutz, organizational democracy, community life, gender and social justice. Email: palgi@yvc.ac.il

Dr Christian Schott is a Senior Lecturer in Tourism Management at the Victoria Management School, Victoria University of Wellington (New Zealand). He holds a PhD in Geography from the University of Exeter (UK) from where he joined Victoria in 2002. He has a long-standing research interest in youth travel with a particular focus on the diverse motivations underlying different types of youth mobilities. Most recently his research has examined the self-directed motivations of young adventurous solo travellers and young volunteer tourists by positioning the motivations and sought outcomes in the context of the pivotal life stage of 'youth'. Christian's other research interests include the environmental sustainability of tourism and tourism distribution channels. Email: christian.schott@vuw.ac.nz

Caroline A. Walsh is a PhD student at the University of Kent, UK. Her research area is disabled divers and volunteer tourism. Caroline has a BSc (Hons) in Environment Science and an MSc in Environmental Conservation. She is a Fellow of the Royal Geographical Society, Zoological Society of London, Royal Society of Arts and the Higher Education Academy. She has been diving internationally for the last 20 years and has travelled extensively. Caroline has been leading her own NGO, AMCAI, since 2001 advocating able bodied and disabled people volunteering together in marine conservation activities and projects. AMCAI works to influence policy and practice alike. Due to her expertise Caroline is an observer on two all-party parliamentary groups for marine and coast and disability respectively. Email: c.a.walsh@btinternet.com

Associate Professor Stephen Wearing works at the University of Technology, Sydney (UTS). He has taught as Visiting Fellow at a number of universities in his 23-year career at UTS, including Wageningen University, Netherlands (11 years); Newcastle University, Australia (10 years) and Macquarie University, Australia (7 years). Dr Wearing has received awards from industry and government for his work in the leisure and tourism fields (2007 Frank Steward Award for his major contributions to the parks and leisure industry and in 1992 from the Costa Rican government for services to youth, conservation and community). Dr Wearing has also served on a number of steering committees for the Sustainable Tourism CRC, WICE and the International Union for Conservation Nature (IUCN) Australasian Chapter. He is a Fellow and Life Member of Parks and Leisure Australasia, and has been editor of the *Australasian Parks and Leisure* journal for 9 years. Email: Stephen.wearing@uts.edu.ac

Dr Eugenia Wickens is Reader in Tourism and Course Leader for the MSc Tourism Development and Management in the School of Sport, Leisure and Travel, Faculty of Enterprise and Innovation, Buckinghamshire New University, UK. She is also an expert adviser in Tourism Management for the Higher Education Quality Evaluation Centre, Riga, Latvia. She has edited special editions for a number of international journals. Her research publication record is extensive, comprising numerous papers in journals, chapters in books and conference proceedings. She is a regular reviewer for several journals and often contributes by invitation to tourism key conference addresses. Email: Eugenia.Wickens@ bucks.ac.uk

Dr Anne Zahra lectures in both tourism and hospitality. Her current research areas include tourism policy, tourism organisations, destination management, volunteer tourism, multi-paradigmatic research methodologies and the ontological and epistemological foundations of tourism and hospitality research. Anne has had a twenty-five-year involvement in volunteering both as a volunteer working with rural and urban poor communities in less developed countries and as an organiser of education development projects for volunteers in Fiji, Tonga, India and the Philippines. She has also co-ordinated AusAid projects in South America though her long-term involvement with Reldev Australia Limited, an NGO registered with AusAid. Email: a.zahra@mngt.waikato.ac.nz

Preface

This book is the product of the ongoing research agenda on volunteer tourism developed by the Volunteer Tourism Research Group which is part of the Association for Tourism and Leisure Education (ATLAS). The main aim of the group is to provide a network for critical discussion and dissemination on volunteering within the tourism and associated (sport, events, leisure) sectors.

The Volunteer Tourism Group was launched at a small (10 attendees) but perfectly formed meeting during the ATLAS annual conference in Brighton, UK in July 2008. However, emails from around the world were received saying that whilst people could not make the meeting they would like to be involved and join the discussion list; a testament to the interest in this new but growing area of tourism research. There are now more than 40 members in the research group; whilst the majority of members are academics, there are a number of PhD and MA students and, in addition, it was pleasing to see a small number of staff from companies engaged in volunteer tourism have also joined the network. At the inaugural meeting (July 2008) the research group agreed upon two outputs (1) an edited book and (2) a symposium.

The Council for Australian University Tourism and Hospitality Education (CAUTHE) also has a Special Interest Group (SIG) on volunteering and tourism, which is chaired by Associate Professor Stephen Wearing and Dr Kevin Lyons. It was agreed to organise a joint ATLAS and CAUTHE symposium in Asia, 2009. The Volunteering and Tourism Symposium: Developing a Research Agenda – Linking Industry and Academia took place 14–15 June 2009 at the James Cook University, Singapore Campus.

Consequently, this edited book is the result of contributions from members of the volunteer tourism research group and participants at the symposium. The book includes a wide range of contributions with examples from around the globe, and whilst some of the chapters begin to broaden the thinking about the boundaries of volunteer tourism, it is evident that there are still areas with little or no discussion. The final chapter seeks to address this and indicates areas of future research; furthermore, the volunteer tourism research group will continue to meet at regular intervals. Information relating to the research group can be found at http://www.atlas-euro.org/.

The organisers of the Sustainable Tourism: Issues, Debates and Challenges Conference invited the ATLAS Volunteer Tourism Research Group to run a special stream – *Travel Philanthropy, Volunteer and Charity Tourism* – at their conference in Crete in April 2010. This was a great opportunity for colleagues to get together again and in total, sixteen papers were accepted. Unfortunately, the conference coincided with the volcanic ash disruptions, and whilst the conference did proceed, many colleagues were unable to attend. The proposed outcome for the special stream is a special issue which is currently being negotiated.

Angela M. Benson
Brighton, August 2010

Acknowledgements

This book is the product of a large number of people and organisations whose combined efforts have made it possible.

The Association for Tourism and Leisure Education (ATLAS) has provided the catalyst for researchers of volunteer tourism to come together. In particular, thanks are given to Leontine Onderwater of ATLAS who continues to support the development of the research group.

I would like to thank the editorial team at Routledge for all their support, throughout the publication process – start to finish.

The first international symposium of volunteer tourism owes much to Associate Professor Stephen Wearing and Dr Kevin Lyons, chairs of the volunteering and tourism specialist interest group (SIG) for the Council for Australian University Tourism and Hospitality Education (CAUTHE) who agreed to organise a joint ATLAS and CAUTHE symposium in Asia, 2009. The securing of funds for keynote speakers would not have been possible without their support and that of the CAUTHE Council.

The symposium in Asia, 2009, would not have taken place without the support of colleagues from the BEST Education Network; they provided the vehicle for the meeting which enabled the symposium to go as a precursor to their own conference, Think Tank IX – 'The Importance of Values in Sustainable Tourism'. In addition, they provided the administrative support for the symposium and Australian colleague Dr Deborah Edwards was invaluable in acting as liaison between ATLAS, CAUTHE and BEST and assisting in the organising of the symposium.

This book would not have been possible without the hard work and continuing support of the participants in the Volunteer Tourism Research Group and the participants at the Volunteering and Tourism Symposium, in particular those that contributed: Zoë Alexander, David B. Arnold, Ali Bakir, Jennifer Kim Lian Chan, Bilge Daldeniz, Tracey J. Dickson, Liam Fee, Simone Grabowski, Mark P. Hampton, Joanne Ingram, Steven Jackson, Jamie Kaminski, Anna Mdee, David Mittenburg, Michal Palgi, Christian Schott, Caroline A. Walsh, Stephen Wearing, Eugenia Wickens and Anne Zahra. Your support and professionalism throughout the process was greatly appreciated.

Special thanks for the support of Eugenia Wickens, Buckinghamshire New University, UK, Marios Sotiriades (TEI of Crete, Greece) and the other organisers of the Sustainable Tourism: Issues, Debates and Challenges Conference who invited the ATLAS volunteer tourism research group to run a special stream (Travel Philanthropy, Volunteer and Charity Tourism) at their conference in Crete, April 2010 enabling the continuation of research on volunteer tourism.

Last but not least, I am grateful for the continued support by management and colleagues at the University of Brighton. I am also fortunate to have a partner, family and friends who continually support my career; for this I am truly indebted.

1 Volunteer tourism

Theory and practice

Angela M. Benson

Introduction

The study areas of both tourism and volunteering have long and established histories, both domestically and internationally. The association between volunteering and travel has its roots in the nineteenth century when missionaries, doctors and teachers travelled to aid others; it is more recent that volunteer tourism has become a global phenomenon with future market predictions indicating growth both in size and value (Mintel, 2008; Tourism Research and Marketing (TRAM), 2008). According to the Lasso Communications survey (in Nestora *et al.*, 2009), 2009 was another year of predicted growth with 62 per cent of volunteer tour operators expecting to send more volunteers abroad than the previous year, 16 per cent sending a similar amount and 20 per cent expecting to send less. At present there are no clear statistics to the size of the volunteer tourism market with the majority of figures being derived from website hits (see Chapters 3 and 14), volunteer surveys and supply side operators. Mintel (2008) estimated that the market reached US$150 million in 2006, TRAM (2008) suggests that 'the total expenditure generated by volunteer tourism is likely to be between £832 million ($1.66 billion) and £1.3 billion ($2.6 billion)' (p.42) 'with a total of 1.6 million volunteer tourists a year' (p.5).

The volunteer tourism sector has seen a proliferation of organisations moving into this market place. Whilst many of the volunteering opportunities are often linked to charitable organisations, it is also evident that some of the growth in this sector is by profit-making companies, and whilst some of these can be linked to social entrepreneurship others are purely commercial. The projects on offer are wide ranging: social, community conservation, ecological health and educational. The marketplace is already becoming segmented with programmes being directed towards individuals, families, groups, students (in particular the gap year students), career breaks and the corporate market. With an ever-growing myriad of pricing structures, for example, organisations are now advertising: free projects (although you have to buy your own flights); discounted projects; and low cost projects, while other organisations just quote a price. This growth and segmentation of the marketplace has produced a range of resources, websites and publications, which are regularly updated, largely descriptive but offer information that outlines the numerous volunteer tourism projects available (for example Ausender and McCloskey, 2008; Hardy, 2004; Heyniger, 2007; Hindle *et al.*, 2007). The message of some of this material has changed over recent years and now includes challenging the ethical status of volunteer tourism rather than the previous passive acceptance of volunteer tourism as a 'saving the world' concept.

The growth of the volunteer tourism product in the marketplace has been accompanied by academic activity and slowly a body of research has emerged. Whilst there were a small number of fragmented articles prior to 2000, the book by Wearing (2001) *Volunteer Tourism: Experiences that make a Difference* seemed to act as the catalyst for the literature that followed. Since then there have been two further academic books (Holmes and Smith, 2009; Lyons and Wearing, 2008) and the publication of this edited book (2010). In 2003, a special edition, featuring eight articles, in the journal *Tourism Recreation Research* was devoted to the subject of volunteer tourism with Stephen Wearing acting as the guest editor. In 2009 another special issue on volunteer tourism edited by Lyons, Wearing and Benson was published by the Annals of Leisure Research, with many of the articles being developed papers from the International Symposium of Volunteering and Tourism that was held in Singapore in June 2009. Within the growing number of journal articles that have been published over the last decade the research has focused heavily on the volunteer (Benson and Siebert, 2009; Brown and Lehto, 2005; Brown and Morrison, 2003; Campbell and Smith, 2006; Galley and Clifton, 2004; McGehee and Santos, 2005; Mustonen, 2005; Stoddart and Rogerson, 2004) and more specifically the self (Wearing, 2002, 2003; Wearing and Deane, 2003; Wearing and Neil, 2000). A number of studies have focused on volunteer tourism and host communities (Broad, 2003; Clifton and Benson, 2006; Gard McGehee and Andereck, 2009; Higgins-Desbiolles, 2003; Singh, 2002, 2004) and in particular cultural aspects (Lyons, 2003; McIntosh and Zahra, 2005; Raymond and Hall, 2008). Wearing *et al.* (2005) examined NGOs as a key stakeholder between volunteer tourists and communities. A small number of papers have engaged in research linked to the organisations and management: Wearing (2004) has concentrated on conservation organisations; Coghlan (2007) examined a number of volunteer tourism organisations and offered an image-based typology; a further study by Coghlan (2008) explored the role of organisation staff. Blackman and Benson (2010) reviewed the role of the psychological contract in managing volunteers. Raymond (2008) studied the role of sending organisations and the extent to which they make a difference and Benson and Henderson's forthcoming article (2011) conducted a strategic analysis of research volunteer tourism organisations. To some extent this edited book mirrors the last decade of literature, in that the majority of the chapters are concentrated on the volunteer, a small number pick up on the theme of the volunteer – host relationship and organisations/management. Clearly, the volunteer remains the focus of current research on volunteer tourism.

Volunteer tourism: theoretical frameworks and practical applications

Reflecting the growth in volunteer tourism, this book provides a collection of chapters about tourism and volunteering that is theoretically rich, practically applied and empirically grounded. The contributions are from a range of international scholars who identify and address complex issues associated with both traditional and contemporary volunteer tourism. The current literature on volunteer tourism has contributors from Europe, Australia and the US; this book echoes this, in that authors from around the globe have contributed.

The book is aimed at two kinds of reader. The first is the academic community, as it is intended to provide a contribution to the literature on volunteer tourism: researchers, academic staff and students at all levels; undergraduates, postgraduates and PhD candidates who have a specific interest in volunteer tourism but also across a broader range of curriculum areas which includes tourism studies and management; development and communities studies; leisure studies, cultural geography, volunteer studies and management, non-profit management and sociology. The second audience is non-academic, which includes practitioners, NGOs, policy makers and governments. To this end, this book acts as an important source to develop knowledge and understanding of this significant social movement particularly in light of volunteer tourism being a worldwide phenomenon with much of the volunteer activity taking place in under-developed and developing countries or countries in transition. It is anticipated that the growing number of practitioners, both commercial and non-commercial operators, that supply the volunteer tourism product in an international marketplace would find this book useful when developing their volunteer management and marketing strategies, as it identifies both theoretical and practical application for their consideration.

The book consists of two parts. Part 1 focuses on the 'volunteer: motivations, experiences and the self' and consists of seven chapters (2–8). All of these chapters are from the perspective of international volunteers taking part in projects in developing countries with case study data collected from around the globe: South Africa (Chapter 2), Nicaragua (Chapter 3), Nepal (Chapter 4), Guatemala (Chapter 5), Sabah, Malaysia (Chapter 6), Pacific and Asia (Chapter 7), Israel (Chapter 8) and mainland Malaysia (Chapters 3 and 9). Chapter 2 by Alexander and Bakir highlights a number of the issues related to understanding the term 'voluntourism', in particular drawing attention to the terms 'volunteer tourism' and 'voluntourism'. As part of the theoretical construct in trying to determine an understanding of the concept of voluntourism they use the voice of the volunteer. The issue of terminology spills over from Chapter 2 into Chapter 3, where Dalendiz and Hampton examine two groups of volunteers – the VOLUNtourists and volunTOURISTS. The analysis of motivations of both groups is discussed and the issues of short-term and long-term volunteers are highlighted. In Chapter 4, Wickens highlights the volunteers' journey of self-discovery through examining motivations, cultural experience and in particular culture shock. Chapter 5 (Schott) continues the theme of motivation by examining the self-development motivations and experiences of a group of youths volunteering at an orphanage in the Guatemalan rainforest.

4   *Angela M. Benson*

Chapter 6 by Chan explores volunteer motivations and experiences by examining four volunteer tourist sites; the findings are used to suggest a developmental and promotional framework in order to better facilitate the management of both the volunteer and the sites. The last two chapters in this part are retrospective studies of volunteers. Zahra in Chapter 7 examines the motivation and experiences of volunteers by asking volunteers to reflect upon their volunteer experience (1989–2000). She draws on the narratives of volunteers in order gain insight into long-term and lasting impacts of the volunteer tourist experience on the participants. Mittenberg and Palgi (Chapter 8) commence by discussing volunteer tourism and contextualising volunteers in the Kibbutz within this framework. The chapter continues by exploring how volunteers evaluate the outcomes of their own Kibbutz volunteer experience by using veteran Kibbutz volunteers from 1970–2000.

Part 2 (Chapters 9–6) is dedicated to 'expanding the boundaries of volunteer tourism research'. Whilst Chapter 9 could easily have been included in Part 1, it is seen as more appropriate here in that Walsh and Hampton examine volunteers within another tourism niche – dive tourism – but more specifically, disabled dive tourism. As outlined in the chapter the issue of volunteering in tourism by disabled people is lacking within academic literature and whilst the sample is small the findings indicate that disabled divers have similar motivations and experiences to their able-bodied peers; however, the constraints to participation are also evident. Jackson in Chapter 10 examines volunteer holiday leaders' values and motives in respect of long-term allegiance to one organisation. The study is contextualised within the National Trust, a UK charitable organisation that manages visitor attractions for an international audience. Kaminski, Arnold and Benson in Chapter 11 discuss that volunteers have been used in archaeological excavations since the 1960s in contexts which are at least in part touristic, where they represent the only practical solution to labour shortage. However, despite this, the literature discussing volunteers in the context of archaeological projects is minimal. The purpose of this theoretical chapter is to set the framework for further discussion. Chapter 12 by Dickson contends that volunteers are an essential element for the effective operation of a range of businesses and activities and as such, the appropriate management of volunteers is paramount. The chapter demonstrates how a risk management process could be applied to the management of volunteers in a tourism context in order to create an enduring legacy of volunteers. The next three chapters are framed within the concept of 'development' albeit by examining different components. Wearing and Grabowski's (Chapter 13) study uses a decommodified approach which aims to explore development through tourism which is defined by local communities in their own terms. They offer the case study of Youth Challenge Australia to contextualise their ideas of the volunteer–host experience. In Chapter 14 Ingram offers a theoretical paper that discusses volunteer tourism through the lens of development theory and asks 'How do we know [volunteer tourism] is making a difference?' This theme of 'making a difference' continues in the next chapter (15) by Fee and Mdee who outline a number of arguments highlighting volunteer tourism's imperfections and a number of possible solutions by using concepts around the process of accreditation. The final chapter (16) draws together key themes from the book, other literature and recent conferences/symposiums on volunteer tourism in an attempt to set a structure for a future research agenda.

References

Ausender, F. and McCloskey, E. (Eds) (2008) *World Volunteers: The World Guide to Humanitarian and Development Volunteering* (4th edn). New York: Universe Publishing.

Benson, A. M. and Henderson, S. (2011) A Strategic Analysis of the Research Volunteer Organisations. *Service Industries Journal,* 31(6).

Benson, A. M. and Siebert, N. (2009) Volunteer Tourism: Motivations of German Participants in South Africa. *Annals of Leisure Research,* 23(3&4): 295–314.

Blackman, D. and Benson, A. M. (2010) Research Volunteer Tourism: The Role of the Psychological Contract in Managing Research Volunteer Tourism. *Journal of Travel and Tourism Marketing,* 27(3): 1–15.

Broad, S. (2003) Living the Thai Life – A Case Study of Volunteer Tourism at the Gibbon Rehabilitation Project, Thailand. *Tourism Recreation Research,* 28(3): 63–77.

Brown, S. and Lehto, X. (2005) Travelling with a Purpose: Understanding the Motives and Benefits of Volunteer Vacationers. *Current Issues in Tourism,* 8(6): 479–496.

Brown, S. and Morrison, A. M. (2003) Expanding Volunteer Vacation Participation: An Exploratory Study on the Mini-Mission Concept. *Tourism Recreation Research,* 28(3): 73–82.

Campbell, L. and Smith, C. (2006) What Makes Them Pay? Values of Volunteer Tourists Working for Sea Turtle Conservation. *Environmental Management,* 38(1): 84–98.

Clifton, J. and Benson, A. M. (2006) Planning for Sustainable Ecotourism: The Case of Research Ecotourism in Developing Country Destinations. *Journal of Sustainable Tourism,* 14(3): 238–254.

Coghlan, A. (2007) Towards an Integrated Image-Based Typology of Volunteer Tourism Organisations. *Journal of Sustainable Tourism,* 15(3): 267–287.

Coghlan, A. (2008) Exploring the role of expedition staff in volunteer tourism. *International Journal of Tourism Research,* 10(2): 183–191.

Galley, G. and Clifton, J. (2004) The Motivational and Demographic Characteristics of Research Ecotourists: Operation Wallacea Volunteers in South-east Sulawesi, Indonesia. *Journal of Ecotourism,* 3(1): 69–82.

Gard McGehee, N. and Andereck, K. (2009).Volunteer Tourism and the 'Voluntoured': The Case of Tijuana, Mexico. *Journal of Sustainable Tourism,* 17(1): 39–51.

Hardy, R. (2004) *The Virgin Guide to Volunteering: Give Your Time and Get Work and Life Experience in Return.* London: Virgin Books.

Heyniger, C. (2007) *The Complete Guide to Volunteer Tourism.* Available from http://matadornetwork.com/bnt//2007/07/23/the-complete-guide-to-volunteer-tourism.

Higgins-Desbiolles, F. (2003) Reconciliation Tourism: Tourism Healing Divided Societies! *Tourism Recreation Research,* 28(3): 35–44.

Hindle, C., Miller, K., Wintle, S. and Cavalieri, N. (2007) *Volunteer: A Traveller's Guide to Making a Difference Around the World.* London: Lonely Planet.

Holmes, K. and Smith, K. (2009). *Managing Volunteers in Tourism: Attractions, Destinations and Events.* Oxford: Butterworth-Heinemann.

Lyons, K. D. (2003) Ambiguities in Volunteer Tourism: A Case Study of Australians Participating in a J-1 Visitor Exchange Programme. *Tourism Recreation Research,* 28(3): 5–13.

Lyons, K. D., and Wearing, S. (Eds) (2008) *Journeys of Discovery in Volunteer Tourism: International Case Study Perspective.* Wallingford, Oxfordshire: CABI

McGehee, N. G. and Santos, C. A. (2005) Social Change, Discourse and Volunteer Tourism. *Annals of Tourism Research,* 32(3): 760–779.

McIntosh, A. J. and Zahra, A. (2005) *Alternative Cultural Experiences Through Volunteer Tourism.* Paper presented at the Association for Tourism and Leisure Education (ATLAS), Catalonia, Spain.

Mintel (2008) *Volunteer Tourism – International – September 2008.* London: Mintel International Group Ltd.

Mustonen, P. (2005) Volunteer Tourism: Postmodern Pilgrimage? *Journal of Tourism and Cultural Change,* 3(3): 160–177.

Nestora, A., Yeung, P. and Calderon, H. (2009) *Volunteer Travel Insights 2009.* Report by Bradt travel guides, Lasso communications and GeckoGo. Available at http://www. geckogo.com/volunteer/report2009.

Raymond, E. (2008) Make a Difference! The Role of Sending Organisations in Volunteer Tourism. In K. D. Lyons and S. Wearing (Eds), *Journeys of Discovery in Volunteer Tourism* (pp. 48–62). Wallingford, Oxon: CAB International.

Raymond, E. M. and Hall, C. M. (2008) The Development of Cross-Cultural (Mis) Understanding Through Volunteer Tourism. *Journal of Sustainable Tourism,* 16(5): 530–543.

Singh, T. V. (2002) Altruistic tourism: Another shade of sustainable tourism: The case of Kanda community. *Tourism: An International Interdisciplinary Journal,* 50(4): 371–381.

Singh, T. V. (2004) *New Horizons in Tourism: Strange Experiences and Stranger Practices.* Wallingford, Oxon: CABI Publishing.

Stoddart, H. and Rogerson, C. M. (2004) Volunteer Tourism: The Case of Habitat for Humanity South Africa. *GeoJournal,* 60(3): 311.

Tourism Research and Marketing (TRAM). (2008) *Volunteer Tourism: A Global Analysis.* ATLAS Publications: Barcelona.

Wearing, S. (2001) *Volunteer Tourism: Experiences that make a Difference.* Wallingford, Oxon: CABI Publishing.

Wearing, S. (2002) Re-centring the Self in Volunteer Tourism. In G. M. S. Dann (Ed.), *The Tourist as a Metaphor of the Social World* (pp. 237–262). Wallingford, Oxon: CABI Publishing.

Wearing, S. (2003) *Volunteer Tourism.* Wallingford, Oxon: CABI Publishing.

Wearing, S. (2004) Examining Best Practice In Volunteer Tourism. In R. A. Stebbins and M. Graham (Eds), *Volunteering as Leisure/Leisure as Volunteering: An International Assessment* (pp. 209–224). Wallingford: CABI Publishing.

Wearing, S. and Deane, B. (2003) Seeking Self: Leisure and Tourism on Common Ground. *World Leisure,* 45(1): 4–12.

Wearing, S., McDonald, M. and Ponting, J. (2005) Building a Decommodified Research Paradigm in Tourism: The contribution of NGOs. *Journal of Sustainable Tourism,* 13(5): 424–439.

Wearing, S. and Neil, J. (2000) Refiguring Self and Identity Through Volunteer Tourism. *Loisir et Societe/Society and Leisure,* 23(2): 389–419.

Part 1

The volunteer

Motivation, experiences and the self

2 Understanding voluntourism

A Glaserian grounded theory study

Zoë Alexander and Ali Bakir

Introduction

Studies on voluntourism tend to focus on tourist motivation (Brown, 2005; Coghlan, 2006a; McGehee and Andereck, 2008; Wearing, 2001); impacts on communities (Wearing and Lee, 2008) and the environment (Pezzullo, 2007); sustainability (McIntosh and Zahra, 2007); tourism marketing (Coghlan, 2007); tourist satisfaction (Brown, 2005; Coghlan, 2006b); and impacts on society generally (McGehee, 2008; Pezzullo, 2007). In the area of volunteer tourism, the academic literature seldom uses the term 'voluntourism'; it does not offer a serious treatment of volunteer tourism and, as such, the concept of voluntourism remains largely undertheorised. Nevertheless, the most commonly used definition of volunteer tourists in this literature is the one advanced by Wearing (2001: 1) as those who 'volunteer in an organized way to undertake holidays that may involve the aiding or alleviating of the material poverty of some groups of society, the restoration of certain environments, or research into aspects of society or environment'.

The volunteer tourism industry, on the other hand, often uses the term 'voluntourism' usually quoting David Clemmons (2009), of Los Niños and voluntourism.org, as 'the conscious, seamlessly integrated combination of voluntary service to a destination and the best, traditional elements of travel – arts, culture, geography, history and recreation – in that destination'. Both definitions seem to combine two elements: tourism[1] and volunteering[2] at the destination visited (Billington, Carter and Kayamba, 2008; Wearing, 2001). The tourism component itself is bedevilled by conceptual weakness and fuzziness (Cooper *et al.*, 2005) due to its very broad nature (Gilbert, 1990) and the expanding spread of activities it covers, including volunteering. Similarly, there is no universal definition of volunteering (Volunteering England, 2008). The word is used by different people and different sectors to mean different things:

> For instance government schemes whereby people have to work for charities in return for benefits are sometimes described as 'voluntary' but many people would argue that since people taking part have to do the work in return for benefits they are not 'volunteers'. Equally many people work unpaid in order to gain experience in very competitive areas like television but most people would not describe them as volunteers.
>
> (Volunteering England, 2008)

Reed (2009) suggests that the definitions of volunteer tourism are too limiting and tell the reader very little about the phenomenon. For example, these definitions do not elaborate the activities undertaken, purpose for travel, time frame involved and potential change in the voluntourist (see Callanan and Thomas, 2005; McIntosh and Zahra, 2008; and Wearing, 2001). As a result, academics are still feeling their way in defining the term and the phenomenon (McGehee, in Sookhan, 2009); this is exacerbated by its broad nature and by the myriad of terms it is associated with such as ecotourism, sustainable tourism, participatory travel, international volunteering, charity tourism, alternative tourism, adventure tourism, responsible tourism and philanthropic travel, cultural tourism and experiential tourism (Wearing, 2001).

Method

Grounded theory is a popular method for collecting and analysing data in a new area of research (Allan, 2003; Bakir and Bakir, 2006; Dick, 2005; Patton, 1990). It was the natural choice for obtaining an understanding of voluntourism because there was no universal definition of the term. The grounded theory methodology provided the tools to resolve this definitional problem and to generate a new explanation. Glaser's (1992) version of grounded theory (refer to Glaser and Strauss, 1967; Glaser 1993; 1998; 2001), with its clearly defined steps, was used and data collected from 67 web-based questionnaires supplemented by testimonials from voluntourists, academic literature, articles, general correspondence, websites, talk shows, blogs and media stories.

The web-based questionnaire was set up as part of a much larger research project looking at the effect of international voluntourism on voluntourists visiting South Africa. The web has been used for questionnaire surveys in voluntourism recently, for example, Seibert and Benson (2009) studied the motivations of 80 German voluntourists who visited South Africa and McGehee, Clemmons and Lee (2009) used a web-based quantitative survey to identify the motivations of more than 1,100 people worldwide. The use of a web-based questionnaire was driven by the dispersed nature of the sample and its size which provided a large amount of data for analysis. The quality of the data is often reflective of the sampling technique which was a volunteer sample[3] in the case of the voluntourists and an opportunity sample[4] for other participants. The website www.bucksresearch.org.uk was specifically designed to be simple but functional. The website was formally user tested by the first author prior to it being launched in the public domain. After testing, a pilot study was conducted on the website using the first ten participants to test out the data collection process and volume test the technical aspects of the website. Participants then linked to the website www.bucksresearch.org.uk to complete the questionnaire, either through their online booking confirmation with AVIVA SA (a voluntourism provider in South Africa) or via an email sent directly from the first author. The web-based questionnaire was short and contained two closed questions (age and gender), two open questions for collecting further demographic information (occupation and country of residence) and some open questions including:

- What do you think you will do on the project?
- What do you want to do with your spare time on your trip?
- What is your understanding of international volunteering (also known as voluntourism or volunteer tourism)?
- Why are you participating in this type of holiday?

The questionnaire was carefully designed to encourage people to complete it in that the questions were simple and unambiguous as possible and measured the objectives of the research. According to Ryan (1995) and Cooper *et al.* (2005), these are important considerations for successful questionnaires. There was an incentive for the participants to complete the questionnaire as part of the larger research project. According to Dommeyer (1988), response rates increase when there is some kind of reward. The completed questionnaires were received automatically into the first author's email account. The responses were printed, coded and filed. The questionnaires were supplemented by data from testimonials, academic literature, articles, general correspondence, websites, talk shows, blogs and media stories.

All data were reviewed, coded, compared and analysed, sentence by sentence, in an ongoing process whose aim was to allow relevant themes to emerge. Figure 2.1 shows a working example of this process.

The emerged themes were categorised, categories were related to each other and along with their properties created the core category. Data were collected until such a time that any new data added nothing to what already had been discovered, where a theoretical saturation point was achieved (Glaser and Strauss 1967). During data collection, any ideas, thoughts, connections, theory or questions that emerged were recorded as memos (Glaser and Strauss, 1967; Glaser, 1998) by way of pocket cards; these were then related to the tourism literature. The analysed data (the categories), along with the memos, were sorted into a sequence such that a structure emerged which could then be described in words (refer to Glaser, 1993, for examples), providing a detailed definition of voluntourism. Glaser's approach was used over Strauss and Corbin's (1990; 1998) for its more intuitive and less complex analysis and coding techniques which involved identifying key points rather than individual words; it was also easier to understand and visualise for this research. Furthermore, grounded theory methodology was preferred over other methodologies because of its common sense approach to 'all is' data. This enabled the incorporation of the perceptions of tourists as well as the indirect experiences of others, such as family members and voluntourism providers, and did not rely exclusively on the experiences of the voluntourists themselves.

Margin Themes	**Text** Codes are underlined
Involvement Awareness Impact on people Travel	Direct <u>involvement</u> and a direct <u>understanding</u> of what is happening there and <u>how your presence is impacting on those people</u> in <u>that part of the world</u>

Figure 2.1 An extract of the coding and analysis process for one piece of data

Findings

The term that emerged as an explanation of voluntourism is 'engagement in volunteer work as a tourist'. As the term indicates, it consists of three main categories: 'engagement', 'tourist' and 'volunteer work'. The accumulated findings are shown diagrammatically in Figures 2.2–2.4.

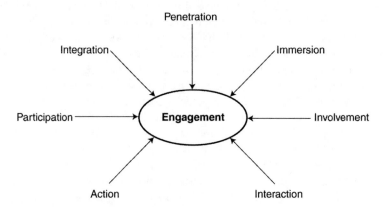

Figure 2.2 The concepts relating to the category of engagement

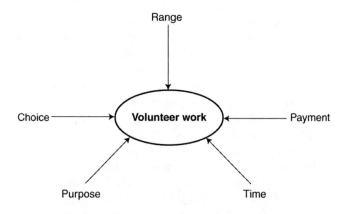

Figure 2.3 The concepts relating to the category of volunteer work

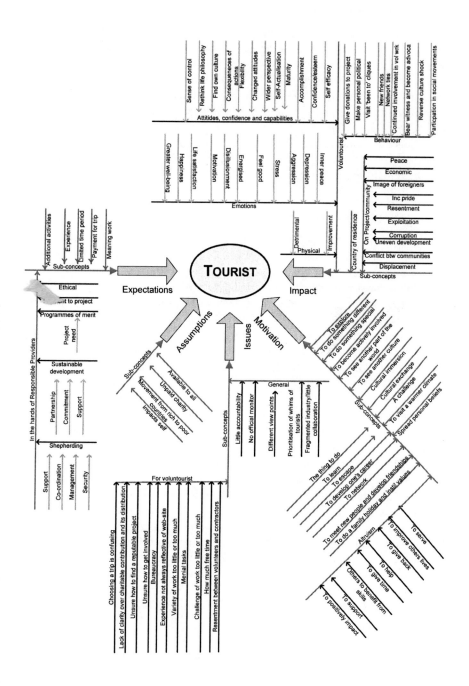

Figure 2.4 The concepts relating to the category of tourist

In these figures, the reader will see the three categories and their constituent concepts/themes. It was through the process of coding/conceptualising and categorising that the data led to the emergent theoretical explanation. 'Engagement' emerged as the core category which subsumed the other two categories of 'volunteer work' and 'tourist', and differentiated voluntourism from other forms of tourism and volunteering. In what follows we will provide a detailed discussion of the engagement core category with less emphasis on the subsumed categories of volunteer work and tourist. We will also apply Glaser's (1998) criteria to determine whether the term 'engagement in volunteer work as a tourist' is trustworthy to be used in the substantive area.

The concept of 'engagement'

Engagement emerged as the essential, key component of voluntourism; it encompassed the concepts of participation, action, integration, penetration, interaction, involvement and immersion (refer to Table 2.1 for descriptions).

A voluntourist responded: 'You go and help where ever you may be needed with various ... things, for example, I helped out with some teaching and then in surrounding villages I put up fences' (Karen, United Kingdom). What this respondent did while volunteering was that she 'acted' and 'interacted'; 'helped', engaged in 'teaching' and 'put[ting] up fences'. For her, voluntourism was a participatory travel experience (Billington *et al.*, 2008; Coghlan, 2006a; McGehee, 2005) which went beyond the need of tourists to gaze, daydream and fantasise, as suggested by Urry (1990), to being active participants wanting to, as another respondent relayed, 'give time to others to improve their lives' (Pat, South Africa). Engagement in voluntary work as a tourist is a new emerging type of tourism, a niche tourism, which contrasts with the concept of 'the gaze', characteristic of the more traditional and homogenous mass tourism (Philbrook, 2007). The 'need' of the voluntourist to be involved, to engage, characterises the modern tourist's desire to experience a place and its culture rather than merely stand back and gaze: 'it is about helping others whilst you enjoy the culture and climate of another country usually hotter than the UK but more disadvantaged' (Skinzzzy, United Kingdom), 'a big piece of it is cultural immersion' (Kimberley, Australia). It is this concept of cultural immersion which prompted some authors to draw parallels between voluntourism and cultural tourism (see, for example, Brown, 2005). To one voluntourist, 'it [voluntourism] was about being able to physically and emotionally immerse oneself in the local culture and community' (Tammy, Canada); a kind of 'integration' and 'penetration' where the tourist seeks a natural and authentic experience (see MacCannell, 1989; and Cohen, 1988) rather than a contrived experience reported of tourism by Boorstin (1961). However, as Kontogeorgopoulos (2003) points out in his study of tourists in Thailand, authenticity means different things to different people; nevertheless, it is an individual's desire for cultural authenticity that differentiates voluntourism from traditional mass tourism, and confers on it the term 'alternative form of tourism' (see Wearing, 2001).

Table 2.1 The core category of engagement with its concepts and properties

Core Category: Engagement	
Concepts	*Property*
Participation	Sharing in an activity.
Action	Doing something.
Integration	Mixing with other people or ethnic groups.
Penetration	To see clearly and deeply.
Interaction	To have an effect on each other.
Involvement	To become connected or associated with.
Immersion	To engross yourself and get absorbed in.

Engagement is a more comprehensive term than that of 'interaction' proposed by Wearing (2001). Many tourism activities involve interaction with the 'other' (Wearing, 2001), which may result in some impacts on the few individuals concerned, but it does not necessarily result in 'connecting' with one another. Engagement, on the other hand, usually involving a larger number of people, is about connecting with the 'other' in some kind of meaningful action. It is associated with Csikszentmihalyi's (1975) 'flow experience' which has been used more recently by Ryan (1996) to describe the role of guides in white-water rafting; flow experience, Csikszentmihalyi posits, is 'complete *involvement* of the actor with his activity' (1975: 36). Geoff Brown, a Community Partnership Coordinator at Florida Fish and Wildlife Conservation Commission, uses the term 'connection' rather than 'interaction' in describing what voluntourists do:

> perhaps I should set up some activities here in Florida to recruit the breed of folks [travellers] who want a 'connection'. They may be interested to find projects like these [litter pick up/ invasive non-native pest-plant removal] during a visit or vacation to help them feel connected to the destination in a way that a 'tourist' would not.
>
> (Geoff Brown, Recreation Services, Florida Fish and Wildlife Conservation)

The engagement themes of connection, commitment and immersion recurred in many respondents' perception and description of voluntourism: '*I will be assisting workers in the children's home and tending to the children's needs, I will be helping the project reach its objectives' (Audrey, Ireland);* 'it is a type of tourism which involves engaging in activities to further a charitable cause' (Ali, United Kingdom).

Although the concept of engagement is found in this study to conceptualise the voluntourism experience, it is nevertheless not peculiar to voluntourism. In the past decade or so, a body of literature has sprung around what has come to be known as engagement theory. Engagement theory has emerged formally in technology-based teaching and learning environments (see Shneiderman, 1994; 1998; Shneiderman *et al.*, 1995; Kearsley, 1997). The idea underlying engagement theory is that students optimise their learning by being meaningfully engaged in activities *through* interaction with others and doing worthwhile tasks (Kearsley and Schneiderman,

1999). The concept itself has also been applied to business environments: in engaging employees to increase job performance and job satisfaction (Maslach and Leiter, 2008); in marketing, engaging consumers (see Spillman, 2006); in conflict negotiation (see Coleman *et al.*, 2008); in tourism planning, stakeholder engagement to create and execute strategy (Cooper *et al.*, 2005); and in psychology, engagement with everyday life to enhance well-being (see Csikszentmihalyi, 1997).

Kearsley and Schneiderman's (1998) engagement theory is predicated on three interrelated components: 'Relate–Create–Donate'; these components are seen to optimise learning through collaborative teamwork and project-based work which have an outside 'authentic' focus. Kearsley and Schneiderman (1998) use the term 'relate' to refer to team efforts that involve communication, planning, management and social skills. This collaborative process, they argue, forces students to clarify and verbalise their problems, thereby facilitating solutions. It also offers an opportunity to work with others from quite different backgrounds, thus facilitating an understanding of diversity and multiple perspectives. The second component, 'create', involves making learning moments creative and purposeful. Kearsley and Shneiderman (1998) see this as a process where students define a project and then concentrate their efforts on applying their ideas in a specific context; not only do students define their own project (even if the topic has been chosen by the instructor), they also have a sense of control over their learning. 'Donate', the third component, stresses the importance of making a contribution to an outside 'customer' (e.g. individuals, community groups, campus organisations, local businesses and government agencies). Kearsley and Shneiderman (1998) point out that this component makes the project 'authentic'. This view of engagement is not very dissimilar to the one found in this study, and emerged as central to understanding voluntourism. Voluntourists 'relate' and collaborate by interacting and integrating with other volunteers and the communities themselves. They 'donate' to an outside customer, the community where the project is located. However, the concept of engagement in voluntourism differs from Kearsley and Shneiderman's (1998) theorisation of engagement in that voluntourists do not 'create' projects to address problems, rather, they 'dedicate' themselves by participating in purposeful work and applying their knowledge and skills to existing worthwhile projects. To draw parallels and contrasts with Kearsley and Shneiderman's engagement components of 'Relate–Create–Donate', this study suggests that engagement in voluntourism is predicated on the components of 'Relate–Dedicate–Donate'. The application of engagement theory to voluntourism offers an opportunity for further research because 'Engagement Theory has only been investigated by a few authors, explained in a small number of articles, and therefore, is not yet clearly defined nor comprehensively theorized' (Hughey, 2002: 4).

The concept of 'volunteer work'

As mentioned above, a summarised account of the concept 'volunteer work', another category of voluntourism, will be presented here. The components of this emerging category included the concepts of choice, range of work, payment (to a provider), period of time and specific purpose (refer to Table 2.2. for descriptions).

Table 2.2 The category of volunteer work with its concepts and properties

Category: Volunteer work	
Concepts	*Property*
Choice	Voluntourists can pick from many projects, voluntourism providers (for profit and non-profit) and destinations.
Range	There are a range of projects available from humanitarian projects (trying to improve the conditions of life for people through health, education, repair, renovation, construction, sustainability) to conservation projects (the protection of animals, plants, land and buildings) and disaster mitigation.
Payment	There is some form of payment involved, from the voluntourist.
Time	The volunteer work is done for a set period of time.
Purpose	The volunteer work serves a purpose for the project, provider and voluntourist.

The concept of 'choice' was significant in its frequent appearance in respondents' narratives; one voluntourist who sent notes about her 'do-it-yourself' volunteer trips stated:

> How, doing what, and where you choose to volunteer is entirely up to you. Do you prefer a specific area of the world? Will you survive without running water and flush toilets? Can you get by in a foreign language? What do you want to do? Teach English? Dig ditches? Restore narrow-gauge train tracks?
>
> For one week? Or one year?
>
> (Anna, Sweden)

Evidently, there is more choice in voluntourism than there is in traditional volunteering. Voluntourists can choose where they want to go and how long they stay (O'Connor, 2008). Voluntourism can be tailored to the individual's interests and knowledge base (Selva, 2008). Voluntourists can head to far-flung parts of the world to build walls, dig fields and care for animals in wildlife sanctuaries, working by day and partying by night (Maxwell, 2006). 'So numerous are the volunteer options', said Michelle Peluso (2007), chief executive of Travelocity, 'that many people are confused by the choice'. Although the concept of choice is central to the definition of traditional volunteering, this choice is between doing volunteering or not, rather than the choice between destination, project, time period and tourism activities, as in voluntourism. The Voluntary Service Overseas (VSO, 2007; 2008) indicated that voluntourists are increasingly approaching them and other similar organisations 'as if it was a holiday' and choosing safe destinations. This choice is made possible because of the 'range of projects' available. In traditional volunteering, although there is a wide range of fields such as health and social care, sport and employer-supported volunteering, projects are limited by their aim to benefit the environment or someone (individuals or groups) other than, or in addition to, close relatives (Volunteering England, 2008). However in voluntourism, especially where the provider is a travel company rather than a non-profit organisation or charity, the projects are travel products (Miller, 2007) and clients 'pay' for the services and in return are given access to an infrastructure and a placement to suit the traveller, taking into consideration 'time' restrictions and 'specific purpose' or requirements. The purpose may therefore extend to satisfying the self (the customer) and therefore voluntourism aims to benefit multiple parties (Clemmons, 2008).

The concept of 'tourist'

The concept of 'tourist', a category of voluntourism, was found to encompass the diverse components of expectations, assumptions, issues, motives and awareness of impacts, reflecting the very broad nature of voluntourism (refer to Figure 2.4).

Responding voluntourists expected to pay the provider, take a trip for a limited period of time, experience something unique and special, as well as do some tourism activities, in addition to the volunteer work (refer to Table 2.3. for descriptions): 'I think trips are usually between two and six weeks long. The voluntourists pay a lump sum which covers their travel expenses, food, accommodation, etc.' (Laura, South Africa).

Furthermore, voluntourists expected to be in the hands of responsible providers and to carry out meaningful volunteer work. Thoits (2008) comments: 'People want purpose and meaning in their lives. One of the ways to find them is to give assistance to the community and feel that they're doing something that matters', yet voluntourists also want to have some fun and do tourism activities on their volunteer vacation: 'I want to do see everything in Cape Town, do some extreme things like bungee jumping, sandboarding, paragliding, safari' (Toeman, Netherlands). With most projects, there is free time to relax, reflect and explore the community and destination. Voluntourism allows the voluntourist to combine good deeds with the best aspects of travel: recreational activities, cultural events, culinary adventures and even shopping and excursions (Globe Aware, 2007a). Voluntourists have certain assumptions and issues about voluntourism (refer to Table 2.4 for descriptions). They assume that voluntourism is available to everyone because websites such as Globe Aware (2007b) state that 'programmes are equally appropriate for the solo traveller to multigenerational family travel, corporate groups and more'; that the volunteer work itself is unpaid; that it involves travel to a less developed country; and that the experience will have an impact on self, the voluntourist.

Table 2.3 The category of tourist with its concept of 'expectation'

Category: Tourist	
Concepts	*Sub-concepts and properties*
Expectations	*Payment* – voluntourists expect to pay the provider.
	Time – voluntourists expect the trip to be for a limited period of time.
	Experience – voluntourists expect the trip to be unique and special, a journey of discovery.
	Additional activities – voluntourists expect to do fun activities additional to volunteer work such as excursions.

	Responsible providers – voluntourists expect the providers to be responsible by:
	• shepherding
	  – providing support
	  – co-ordinating
	  – providing management
	  – providing security
	• being ethical
	• developing programmes of merit
	  – satisfying project needs
	• making payments to the project
	• ensuring a sustainable development
	  – creating a partnership
	  – being committed
	  – providing support.

	Meaningful volunteer work – voluntourists expect the providers to create meaningful work which: • use their skills • is of benefit • provides contact with the locals • provides authentic experience • provides recognition • provides support • provides respect.

Table 2.4 The category of tourist with its concepts of 'assumptions' and 'issues'

Category: Tourist	
Concepts	Sub-concepts and properties
Assumptions	*Available to all* – voluntourists believe that anyone can participate including the young (16+), gap yearlings, students, postgrads, retirees, professionals, families, corporate groups, middle-aged people, mature people, regular volunteers, occasional volunteers, solos, couples, groups.
	Volunteer work is unpaid charity.
	Movement of volunteers from rich west to poorer developing countries.
	There will be an impact on self – a long lasting change.
Issues	*General*
	• little accountability
	• no official monitor
	• different points of view
	• prioritisation of the whims of tourists
	• industry is very fragmented with little collaboration.
	For the voluntourist
	• choosing a trip is confusing – there is a myriad of pricing structures such as discounted projects, free projects but flights excluded, low cost etc.
	• lack of clarity over charitable contribution to the project and how it is distributed
	• unsure how to find a reputable project
	• unsure how to get involved
	• bureaucracy – such as reference checks and CRB checks
	• experience not always reflective of website advert
	• variety of work – there can be too little or too much variety of work
	• menial tasks
	• challenge of work – there may be too much or too little of a challenge
	• free time – voluntourists are unsure of how much free time they will have
	• resentment between volunteers and 'western' contractors.

A respondent stated: 'It is about helping others and growing from the experience' (Duncan, South Africa). Voluntourists were also concerned with certain issues such as the confusion around finding a reputable trip and what work would be required: 'Weeding through the myriad of volunteer options can be daunting, it was for me' (Anna, Sweden). Voluntourists had numerous motives for participating in voluntourism (refer to Table 2.5 for descriptions). These range from 'spreading personal beliefs', 'doing something different', 'networking', 'visiting a warmer climate', 'experiencing another culture' through to altruism and 'the desire to give back' and 'help others'. One voluntourist who travelled to South

Table 2.5 The category of tourist with its concept of 'motivation'

Category: Tourist	
Concepts	*Sub-concepts and properties*

Motivation	• to do something different
	• to explore
	• to do something special
	• to become actively involved
	• to see another part of the world
	• to see another culture
	• cultural immersion
	• cultural exchange
	• a challenge
	• the thing to do
	• to learn
	• to escape
	• to visit a warmer climate
	• to develop one's career
	• to network
	• to meet new people and develop friendships
	• to go on a family holiday and instil values
	• to spread personal beliefs – primarily religious
	• altruism
	   – wanting to serve others
	   – to improve other people's lives
	   – to give back
	   – to help
	   – to give time
	   – to allow others to benefit from one's skills
	   – to support
	   – to positively impact.

Africa said, 'God called me to do it' (Mark, Ireland), and another person said, 'It is the desire to do something good while at the same time experiencing new places and challenges in locales one might otherwise not visit' (Ali, United Kingdom).

The experience brought about by 'engagement' left noticeable impacts on the voluntourists: 'It hopefully benefits those receiving help and, for the person taking part, I would say it is a wonderful and rewarding experience' (Em, United Kingdom); 'it is a unique travel experience that promotes self-growth while giving something back to wildlife, conservation and community projects in need' (Ed, South Africa); and 'It is more about changing the person that is going on the trip than it is about actually helping the world' (Starlagurl, Canada). The impact on self is supported by research into traditional volunteer work and well-being, showing that people who volunteered had greater well-being overall (Thoits and Hewitt, 2001). Research into voluntourism also shows that voluntourism profoundly affects even the most day-to-day elements of the lives of its participants (McGehee, 2005); one respondent concurred:

Since my involvement in voluntourism and my international travel, I'm profoundly affected by the want and misuse of space that I see. It would be
really easy just to focus on how bad everything is but I try not to do that.
When I came back from Beijing, I sold my house and bought a much smaller
house, and the kitchen is quite small, we call it our European kitchen. You
definitely have to do the butt dance in the kitchen. And I think I am much
more aware of the impact that we have with overspending and the overheat-
ing and over-everything. I recycle and reuse and think about things before I
buy them ... I could go on and *on* ...

(Maude, United States)

There was evidence that voluntourism positively impacts the projects and community too (refer to Table 2.6 for descriptions): 'some [voluntourists] come away
as passionate advocates and long-term donors of the communities they have interacted with, also with heightened awareness of how our actions have consequences
on people even thousands of miles away' (Chris, United Kingdom); and

While at times during my brief three-day sojourn at Mirror I felt inadequate,
useless and under-prepared, there was one moment when my skills came to
the fore, and finally I understood the value of my contribution. A group of
about 20 volunteers, including myself and Paul, some Japanese students and
a bunch of local Akha scholarship kids, headed into the Chiang Rai night
market to hand out leaflets to western tourists, pointing out the dangers associated with giving money to child beggars.

(Julie, Australia)

Table 2.6 The category of tourist with its concept of 'impact'

Category: Tourist	
Concept	*Sub-concepts and properties*
Impact	On project/community • economic benefits • cultivate peace • change image of foreigners • conflict between communities • displacement • increased pride • resentment • exploitation • local corruption • uneven development due to popularity of some projects more than others.

Category: *Tourist*

Concept	Sub-concepts and properties

On voluntourist
- Behaviour
  - give donations to the project
  - make the personal political
  - visit 'been-to' cliques
  - network ties
  - new friends
  - continued involvement in volunteer work
  - bear witness and become advocates
  - reverse culture shock – difficulty finding/resuming work or study, getting used to the pace of life, social difficulty, loneliness, depression
  - participation in social movements at home.

- Attitudes, confidence and capabilities
  - self efficacy
  - increased confidence and esteem
  - sense of accomplishment
  - greater emotional and intellectual maturity
  - self-actualisation
  - wider global perspectives
  - transformed attitudes
  - greater flexibility
  - greater appreciation for the consequences of human action on the environment and communities
  - find own culture
  - rethink life philosophy
  - sense of control over life.
- Emotions
  - inner peace
  - depression
  - aggression
  - different levels of stress
  - feel good for doing more
  - energised to do more
  - anger
  - disillusionment
  - more motivated
  - greater satisfaction with life
  - greater well-being
  - happiness.
- Physical
  - improvement in health
  - detrimental to health.

On country of residence
- resources diverted elsewhere.

However, there are instances of exploitation if projects are not carefully planned and managed (Archer and Cooper, 1994; Theobald, 1994):

> It was disheartening to have travelled 5,000 miles away only to find that all I'm really here for now is money and assets. Everything we give to the children is snatched and more demanded. Should a child receive a new pencil, their friend will want one. There is aggression and so many kids will fight, lie and haggle to get what they want, be it fruit, chocolate or stationery
>
> (Tom, United Kingdom)

and 'I visited one school in Malawi where the head teacher said she took Western volunteers because they were cheaper than paying local staff' (Kate, United Kingdom).

Conclusion

The purpose of this study was to understand voluntourism using Glaser's grounded theory. The theoretical explanation that emerged from the data holds up to Glaser's (1998) criteria for judging the adequacy of the emerging explanation: fitness to the situation, workability and modifiability. The explanation and the concepts that gave rise to it were grounded in the experiences of voluntourists, voluntourism providers and other stakeholders, the perceptions of tourists and research into the different aspects of voluntourism such as motives, experiences and impacts; it thus 'fits' the situation. It also 'works' because it helps people make sense of their experience and to manage the situation better. For example, one voluntourism provider reviewed its web-based project sheets to include the voluntourist's financial contribution to the project. There will be further data collected in the future so that the theory can be improved upon and amended as situations change. In this respect, it will be readily 'modifiable' as new data emerges.

What has so far emerged to describe voluntourism is the key category of 'engagement', necessarily associated with two other emerging categories of 'volunteer work' and 'tourist'. These categories together provided the encompassing explanation of voluntourism as 'engagement in volunteer work as a tourist', pointing to a purposeful connection to particular peoples and places. The categories and concepts generated from this research have highlighted diverse concepts within voluntourism and in this respect have also highlighted the limitations of the research; it is general and does not focus on a specific area in sufficient detail. Furthermore, the first author played a major role in arriving at this explanation as the concepts were her own, emerged as a result of her systematic searching for themes within the data. We feel that this subjectivity limitation was addressed by the extensive data collected and the visibility and audit ability of the detailed process in arriving at the emerged theoretical explanation. Many quotations were used in the chapter, giving the readers the opportunity to interpret the data themselves following our trail, and hopefully arrive at a similar explanation, and by doing so confirming the trustworthiness of the findings.

Glaser's grounded theory offers a definition for voluntourism and, at the same time, allows its deconstruction into constituent elements, offering a contribution to knowledge. It provides a theorisation of the concept of 'touristic engagement', previously undertheorised, thus filling a gap in the tourism literature.

Notes

1 Tourism: Both World Trade Organization and United Nations Statistics Division, 1994 define the concept as 'The activities of persons travelling to and staying in places outside their usual environment for not more than one consecutive year for leisure, business and other purposes' (Cooper *et al.*, 2005: 13).
2 Volunteering: The definition of volunteering used by Volunteering England (2008) is any activity which involves spending time, unpaid, doing something which aims to benefit the environment or someone (individuals or groups) other than, or in addition to, close relatives. Central to this definition is the fact that volunteering must be a choice freely made by each individual. This can include formal activity undertaken through public, private and voluntary organisations as well as informal community participation.
3 A volunteer sample is defined as a sample of people who put themselves forward to take part in the research (Albery *et al.*, 2004).
4 In contrast to a volunteer sample, an opportunity sample is a sample of people willing to take part in the research when asked rather than volunteering to participate (Albery *et al.*, 2004).

References

Albery, I., Chandler, C., Field, A., Jones, D., Messer, D., Moore, S. and Sterling, C. (2004) *Complete Psychology*, Hodder Arnold: Dubai.

Allan, G. (2003) A critique of using grounded theory as a research method, *Electronic Journal of Business Research Methods*, 2(1): 1–10.

Archer, B. and Cooper, C. (1994) The positive and negative impacts of tourism. In W. Theobald (ed.) *Global Tourism: The Next Decade*, Oxford: Butterworth-Heinemann, pp.73–91.

Bakir, A. and Bakir, V. (2006) Dialectical inquiry: A critique of the capacity of Strauss' grounded theory for prediction, change and control in organisational strategy via a grounded theorisation of leisure and cultural strategy. *The Qualitative Report*, 11(4): 687–718. Accessed 20 May 2008, http://www.nova.edu/ssss/QR/QR11-4/bakir.pdf.

Billington, R., Carter, N. and Kayamba, L. (2008) The practical application of sustainable tourism development principles: a case study of creating innovative place-making tourism strategies, *Tourism and Hospitality Research*, 8: 37–43.

Boorstin, D. (1961) *The Image: A Guide to Pseudo-events in America*, New York: First Vintage Books.

Brown, S. (2005) Understanding the motives and benefits of voluntourists: What makes them tick?, *VolunTourist*. Accessed 20 May 2008, http://www.voluntourism.org/news-studyand research1005.htm.

Callanan, M. and Thomas, S. (2005) Volunteer tourism: deconstructing volunteer activities within a dynamic environment. In M. Novelli (ed.) *Niche Tourism: Contemporary Issues, Trends, and Cases*, Oxford: Elsevier Butterworth-Heinemann, pp.183–200

Clemmons, D. (2008) So you may know – gathering the stakeholders, *VolunTourist*, 4(2). Accessed 9 July 2008, http://www.voluntourism.org/news-soyouknow42.htm.

Clemmons, D. (2009) *VolunTourism* international website. Accessed 25 June 2009, http://www.voluntourism.org/.

Coghlan, A. (2006a) Choosing your conservation-based volunteer tourism market segment with care – Part 1: the motivators and the elements of the experience that was most important for the two primary markets, *VolunTourist*. Accessed 20 May 2008, http://www.voluntourism.org/news-studyandresearch0106.htm.

Coghlan, A. (2006b) Choosing your conservation-based volunteer tourism market segment with care – Part 2: ensuring volunteer tourist satisfaction, *VolunTourist*. Accessed 20 May 2008, http://www.voluntourism.org/news-studyandresearch0206.htm.

Coghlan, A. (2007) Towards an integrated image-based typology of volunteer tourism organisations, *Journal of Sustainable Tourism*, 15(3): 267–287.

Cohen, E. (1988) Traditions in the qualitative sociology of Tourism, *Annals of Tourism Research*, 15(1): 29–46.

Coleman, P., Hacking, A., Stover, M., Fisher-Yoshida, B. and Nowak, A. (2008) Reconstructing ripeness I: A study of constructive engagement in protracted social conflicts, *Conflict Resolution Quarterly*, 26(1).

Cooper, C., Fletcher, J., Gilbert, D. and Wanhill, S. (2005) *Tourism Principles and Practice*, 3rd edn, Harlow: Addison Wesley Longman.

Csikszentmihalyi, M. (1975) Beyond boredom and anxiety. In S. Williams (2004) *Tourism: Critical Concepts in the Social Sciences*, London: Routledge.

Csikszentmihalyi, M. (1997) *Finding Flow: The Psychology of Engagement with Everyday Life*, New York: Perseus Book Group.

Dick, B. (2005) Resouce Papers in Action: Grounded Theory a thumbnail sketch, Accessed 14 May 2008, http://www.scu.edu.au/schools/gcm/ar/arp/grounded.html.

Dommeyer, C. (1988) How form of the monetary incentive affects mail survey response, *Journal of the Market Research Society*, 30(3): 379–386.

Gilbert, D. (1990) Conceptual issues in the meaning of tourism. In C. Cooper (ed.) *Progress in Tourism, Recreation and Hospitality Management*, London: Belhaven Press.

Glaser, B. (1992) *Basics of Grounded Theory Analysis. Emergence vs. Forcing*, Mill Valley, CA: Sociology Press.

Glaser, B. (1993) *Examples of Grounded Theory: A Reader*, Mill Valley, CA: Sociology Press.

Glaser, B. (1998) *Doing grounded theory: issues and discussions*, Mill Valley, CA: Sociology Press.

Glaser, B. (2001) *The Grounded Theory Perspective: Conceptualization Contrasted with Description*, Mill Valley, CA: Sociology Press.

Glaser, B. and Strauss, A. (1967) *Discovery of Grounded Theory: Strategies for Qualitative Research* (Reviewed), Aldine Transaction: New Brunswick, NJ and London.

Globe Aware (2007a) Press release: do some volunteer work on vacation. Accessed 28 July 2008, http://www.dallasnews.com/sharedcontent/dws/fea/travel/thisweek/stories/DN-giftofyou_1223tra.ART.State.Edition1.4e2e212.html.

Globe Aware (2007b) Spend your vacation giving back while immersing yourself in another culture. Accessed 28 July 2008, http://www.goabroad.com/providers/globe-aware/programs/spend-your-vacation-giving-back-while-immersing-yourself-in-another-culture-58867.

Hughey, L. (2002) *A pilot study investigating visual methods of measuring engagement during e-learning.* Unpublished report. The Centre of Applied Research in Educational Technologies (CARET), University of Cambridge, UK.

Kearsley, G. (1997) *The Virtual Professor: A Personal Case Study.* Accessed 10 September 2008, http://www.buscalegis.ccj.ufsc.br/revistas/index.php/buscalegis/article/viewFile/3089/2660.

Kearsley, G. and Schneiderman, B. (1998) Engagement theory: a framework for technology-based teaching and learning, *Educational Technology,* 38(5): 20–23.

Kearsley, G. and Schneiderman, B. (1999) Engagement theory: a framework for technology-based teaching and learning. Accessed 10 September 2008, http://home.sprynet.com/~gkearsley/engage.htm.

Kontogeorgopoulos, N. (2003) Keeping up with the Joneses: tourists, travellers, and the quest for cultural authenticity in southern Thailand, *Tourism Studies,* 3(2): 171–203.

MacCannell, D. (1989) *The Tourist: A New Theory of the Leisure Class,* 2nd edn, London: Macmillan Press.

Maslach, C. and Leiter, M. (2008) Early predictors of job burnout and engagement, *Journal of Applied Psychology,* 93, 498–512.

Maxwell, K. (2006) New words on vacation: holidays with a difference, *MED Magazine: The monthly webzine of the Macmillan English Dictionaries,* 40 (July), Macmillan Publishers Ltd, London. Accessed 18 July 2008, http://www.macmillandictionaries.com/med-magazine/July2006/40-New-Word.htm.

McGehee, N. (2005) Voluntourism: making the personal political, *VolunTourist.* Accessed 20 May 2008, http://www.voluntourism.org/news-studyandresearch0805.htm.

McGehee, N. (2008) Study and Research – Initial thoughts and reflections: on the VolunTourist survey, *VolunTourist,* 4(2). Accessed 9 July 2008, http://www.voluntourism.org/news-studyandresearch42.htm.

McGehee, N. and Andereck, K. (2008) Pettin the critters: exploring the complex relationship between volunteers and the voluntoured in McDowell County West Virginina, USA, and Tijuana, Mexico. In K. Lyons and S. Wearing (eds) *Journeys of discovery in volunteer tourism: International case study perspectives,* London: CABI Publishers.

McGehee, N., Clemmons, D. and Lee, Y. (2009) 2008 *Voluntourism Survey Report,* October 2009, VolunTourist, San Diego, CA.

McIntosh, A. and Zahra, A. (2007) A cultural encounter through volunteer tourism: towards the ideals of sustainable tourism? *Journal of Sustainable Tourism,* 15(5): 541–556.

McIntosh, A. and Zahra, A. (2008) Journeys for experience: the experiences of volunteer tourists in an indigenous community in a developed nation – a case study of New Zealand. In K. Lyons and S. Wearing (eds) *Journeys of Discovery in Volunteer Tourism: International Case Study Perspectives,* London: CABI Publishers.

Miller, J. (2007) Voluntourism in Thailand, *Travel Review from Travel Intelligence.* Accessed 2 October 2007, www.travelintelligence.net/wsd/articles/art_1003781.html.

O'Connor, C. (2008) *Voluntourism: The Right Fit.* Accessed 9 July 2008, http://redwoodage.com/content/view/134604/49.

Patton, M. (1990) *Qualitative Evaluation and Research Methods*, London: Sage Publications Ltd.

Peluso, M. (2007) *Voluntourism: Travel that also includes volunteer work*. Accessed 6 May 2008, http://www.wordspy.com/words/voluntourism.asp.

Pezzullo, P. (2007) *Toxic Tourism: Rhetorics of Travel, Pollution, and Environmental Justice*, Alabama: University of Alabama Press.

Philbrook, B. (2007) Voluntourism on the rise, *Home and Away Magazine*, Issue December.

Reed, K. (2009) *The Social Construction of the Volunteer Tourist Identity: Preliminary Analysis of Effects and Implications for Information Behaviours*, Best Education Network Think Tank – on the importance of values in Sustainable Tourism: First Symposium on volunteering and tourism, Conference Paper, Singapore

Ryan, C. (1995) *Researching Tourist Satisfaction. Issues, Concepts, Problems*, London: Routledge.

Ryan, C. (1996) *The Tourist Experience*, London: Continuum.

Seibert, N. and Benson, A. (2009) *The Motivations of German Volunteers: The Case of South Africa*, Best Education Network Think Tank – on the importance of values in Sustainable Tourism: First Symposium on volunteering and tourism, Conference Paper, Singapore.

Selva, M. (2008) Supply Chain – Myanmar & Southeast Asian VolunTourism, *VolunTourist*, 4(2). Accessed 9 July 2008, http://www.voluntourism.org/news-supplychain42.htm.

Shneiderman, B. (1994) *Education by Engagament and Construction: Can Distance Education be Bettter than Face-to-Face?* Accessed 10 September 2008, http://www.hitl.washington.edu/scivw/EVE/distance.html

Shneiderman, B., Alavi, M., Norman, K. and Borkowski, E. (1995) Windows of opportunity in electroni classrooms, *Communications of the ACM*, 38(11): 19–24.

Sookhan, H. (2009) *Research, Volunteer Tourism: A tale of two communities*, Virginia: VirginiaTech University.

Spillman, M. (2006) Cracking the Engagement Code, *iMedia Connection*. Accessed 10 September 2008, http://www.imediaconnection.com/content/10518.asp.

Strauss, A. and Corbin, J. (1990) *Basics of Qualitative Research: Grounded Theory Procedures and Techniques*, California: Sage.

Strauss, A. and Corbin, J. (1998) *Basics of Qualitative Research: Grounded Theory Procedures and Techniques* (2nd edn), California: Sage.

Theobald, W. (1994) *Global Tourism: The Next Decade*, Oxford: Butterworth Heinemann Ltd.

Thoits, P. (2008) Helping and your health. In J. Jacobs, The value of voluntourism, *Creative Living Magazine*, July 2008.

Thoits, P. and Hewitt, L. (2001) Volunteer work and well-being, *Journal of Health and Social Behaviour*, 42: 115–131, American Sociological Association, USA.

Urry, J. (1990) *The Tourist Gaze: Leisure and Travel in Contemporary Societies*, London: Sage.

Volunteering England (2008) *Definition of Volunteering*. Accessed 11 May 2008, http://www.volunteering.org.uk/NR/rdonlyres/4C135BDF-E1E2-43D4-8FD8-DB16AE4536AA/0/DefinitionsofVolunteeringVE08.pdf.

VSO: Voluntary Service Overseas (2007) 'Volunteering tourism puts development charity at risk'. Accessed 8 May 2008, http://www.wcva.org.uk/news/dsp_news.cfm?newsid=968&display_sitedeptid=21.

VSO: Voluntary Service Overseas (2008) *Volunteering*. Accessed 9 May 2008, http://www.vso.org.uk/volunteering.

Wearing, S. (2001) *Volunteer Tourism: Experiences That Make a Difference*. Oxford: CABI Publishing.

Wearing, S. and Lee, D. (2008) Pro-poor tourism: who benefits? Perspectives on tourism and poverty reduction, *Annals of Tourism Research*, 35(2): 616–618.

World Trade Organization and United Nations Statistics Division (1994) *Recommendations on Tourism Statistics*, World Trade Organization, Madrid and United Nations: New York.

3 VOLUNtourists versus volunTOURISTS

A true dichotomy or merely a differing perception?

Bilge Daldeniz and Mark P. Hampton

Introduction

Students and early career professionals from developed countries enter destination countries in the less developed world on a tourist visa to work as volunteers on volunteer placements with non-profit organisations that work on environmental or social projects. This was initially termed volunteer tourism, before the term was shortened to 'voluntourism' and defined as:

> those tourists who, for various reasons, volunteer in an organized way to undertake holidays that may involve the aiding or alleviating the material poverty of some groups in society, the restoration of certain environments, or research into aspects of society or environment.
>
> (Wearing, 2002: 240)

The volunteer tourism literature has so far principally examined short-term volunteer tourism activities, typically designed as breaks of a number of weeks' work on a charitable, social or environmental project in a less developed country, largely targeted at young adults from the developed world and promoted as alternative to a classic backpacker trip (Brown and Morrison, 2003; Stoddart and Rogerson, 2004; Wearing, 2004; McGehee and Santos, 2005; Mustonen, 2005). In this chapter however, it is argued that based on its definition a broader understanding of volunteer tourism can be achieved by incorporating time variations and lifestyle choices of individuals who go beyond the short charitable project work and indeed volunteer for longer periods of time and even without a clear altruistic agenda – although the results may still contribute to poverty alleviation or an increase in environmental awareness.

Two fieldwork case studies were conducted amongst long-term volunteers during 2009. One examined the motivation, lifestyle and living conditions of long-term volunteers in a rural development project in Nicaragua. This group will be termed 'VOLUNtourists' as it is assumed their main motivation for undertaking a longer term commitment in a development project is based on altruistic motives to help the local communities in question. The other case study focused on volunteers in the tourism industry itself. Besides volunteer hotel managers in

Nicaragua, scuba diving professionals in Malaysia and their motivation, living conditions, lifestyle and broad socio-economic and environmental impacts were studied. This group will be referred to as 'volunTOURISTS' for the purpose of this paper, as it is assumed that the tourism activity itself is the main driver for them to accept a long-term volunteer work situation. Furthermore, advertisements in the *Caretaker Gazette*, an internet publication on volunteer caretaker jobs around the world, were analysed for content, notably type of positions offered, the remuneration or compensation entailed and the advertised duration of volunteering.

Literature review

According to Tomazos and Butler (2008), the origins of volunteering abroad seem to go back to projects set up by organisations such as the Service Civil International in Europe in the 1920s and the International Voluntary Service and Peace Corps activities of the United States after the Second World War (Brown and Morrison, 2003). It was not until the 1990s however that volunteering abroad started to become a popular and, now, a mass activity. This has to do with the increasing popularity of the gap year, i.e. young adults taking a break between high school and university, and with reciprocal effect the growth in organisations and now companies who offer gap year activities. In the US, it can also witness the development of the 'alternative' spring break with organisations and companies offering volunteering projects abroad as an alternative to the traditional party-centred spring break.

For immigration purposes the majority of volunteers will declare tourism as the reason for entering a country to avoid work permit issues. This makes it nearly impossible to indicate the current volume of volunteer tourism (Tomazos and Butler, 2008). In August 2009 over six million hits were registered on an internet search portal for the phrase 'volunteer projects abroad', which gives us a clear confirmation of its popularity.

The expectations of participants and their motivation to embark on volunteer tourism activities have been of some interest to researchers. Wearing (2004: 215) states that by understanding the images and attitudes of volunteer tourists 'we may be able to address the managerial implications for organisations operating in the realm of volunteer tourism'.

Motivational factors have traditionally been separated into 'push' and 'pull' factors (Coghlan, 2007). As 'push' factors we can list altruistic motives such as the desire to travel with a purpose, helping to improve the environment, working with communities and in less developed countries. Non-altruistic motives to volunteer abroad are self-enhancement, enhancement of one's curriculum vitae (Callanan and Thomas, 2005), the sensations of accomplishment and belonging, and the desire for social interaction (Gilmour and Saunders, 1995). In addition to this, the development of personal skills and the increase of knowledge, self-confidence and independence were also found to be important to many voluntourists (Wearing, 2004; Webb, 2002).

As 'pull' factors we think of an individual's desire to travel and explore other parts of the world. Gunn (1988) described the importance of images used in tourism destination marketing and their power to create perceived images within the potential tourists. Coghlan (2007) showed that the volunteers were heavily influenced in their decision-making by the images of destinations portrayed in the promotional materials.

Methodology

The three sources

The study was based on three sources: a case study within a rural development NGO in Nicaragua, a study on dive tourism in Malaysia and the analysis of an online publication for job advertisements, *The Caretaker Gazette*.

The case study in Nicaragua was based on participant observation, with one of the authors spending several weeks with a rural development NGO that works mainly with international volunteers. Informal qualitative interviews with the individual volunteers were carried out (n=16 out of a total of 20 current volunteers) and several group discussions developed around the subjects of remuneration, housing, 'visa runs' and travels linked to that, whilst the researcher was present. Prior to undertaking this study, complete anonymity was guaranteed to the organisation and the individuals. Further qualitative interviews were carried out with two volunteer managers of a hotel.

The case study in Malaysia was part of a wider project on the socio-economic and environmental impacts of scuba diving on local communities. Qualitative interviews (n=19) were carried out with international volunteer dive professionals, with dive instructors, dive masters and dive master trainees.

Advertisements in the *Caretaker Gazette*, an internet publication of job offers in the tourism and hospitality as well as social and animal care sector, were analysed for their content, with the main focus being on volunteer posts in tourism and hospitality.

Methods

The different research techniques for both fieldwork parts were chosen based on the requirements of each group of individuals to be studied. Standard qualitative semi-structured interviews were considered appropriate for the dive professionals, who can be described as easy to talk to and often available to sit down for a long enough period of time, at least in the off season, between their diving commitments. The interviews were conducted in the dive shops, which acted as central hubs for the dive professionals to return to between dives or teaching sessions.

The volunteers of the NGO, however, were more difficult to approach. A certain level of trust was needed for them to talk freely about some of the aspects in question, particularly their motivations behind the volunteering, as will become clear when this is discussed later. In logistical terms, their operation was more spread out, with volunteers being based in different remote rural communities, an office building, a workshop and several separate living quarters spread out across a whole suburb in the nearest town. The volunteers would move around between the locations, often in groups, depending on the work they were carrying out, which would have made it impossible to schedule longer interviews. The logistics and above all the level of trust needed made the participation of the researcher in the project essential to obtain the data.

The *Caretaker Gazette*, a piece of 'grey' literature, was consulted as it constitutes a popular tool for the advertisement of volunteer jobs in tourism and hospitality. It underlines the growth of an offer of longer term volunteer jobs within the sector itself.

Survey details and living conditions

A profile of VOLUNtourists and volunTOURISTS is offered in Table 3.1; further details are now discussed.

At the time of the study in Nicaragua, the organisation had twenty volunteers from four different countries, of whom sixteen were interviewed for this research. Excluding the researcher in this count there were eleven volunteers from France, seven from the USA, one from the UK and one from Australia. They were all aged between 23 and 38 and without exception had at least a first university degree. The minimum length of stay was three months, but typically a volunteer would stay for six to twelve months. Two of the volunteers had already extended their stay for another year and two were there on a repeat stay and had been working for the organisation remotely from their home countries throughout several years in the meantime.

The volunteers were funded from varying sources. The French volunteers used their governmental VSI scheme (voluntariat solidarité international), which allows them to take time away from paid employment to volunteer overseas, whilst their pension and social security contributions are paid for by the French government. The scheme furthermore regulates the compensation the organisation has to pay to the volunteer with a minimum amount of €154 per month in the first year to be paid to the volunteer. The rest of the volunteers were self-funded. The NGO has a special section for the volunteers on its webpage that enables future and current volunteers to advertise for funds. They are encouraged to fundraise to pay for their transport to and from the country and the journeys required for 'visa runs' during their stay every three months. As part of the Central America Border Control Agreement (C-4), the volunteers have to leave the C-4 area, which comprises El Salvador, Guatemala, Honduras and Nicaragua, and re-enter in order to obtain a new visa.

Table 3.1 Profiling the volunteers

Aspect	VOLUNtourists – those in development project	VolunTOURISTS – those in diving and hospitality
Legal status in country	• on tourist visa • need for visa runs	• on tourist visa • need for visa runs
Costs	• travel to country • costs for visa runs • nominal rent and food	• travel to country • costs for visa runs • rent and food
Duration	• minimum 3 months, most 6–12 months	• minimum 3 months, most 6 months per location

For their costs of living the volunteers are required to pay a small rent and food contribution, which, in the case of the French volunteers, uses up almost the entire token remuneration the organisation is obliged to pay them under the VSI scheme. Their living conditions are basic, with shared rooms in local standard housing, cold water and frequent power cuts. The organisation employs local staff for cleaning and as kitchen personnel who are in charge of cooking basic communal lunches. Breakfast and dinner are prepared by the volunteers themselves in their own accommodation. The volunteers have four weeks holiday per year, one each for Christmas and Easter and two further to be taken in agreement with the NGO director. The volunteers have to plan them around the necessary 'visa runs'. The volunteers have Saturdays and Sundays off, but most visits to the remote communities are arranged to comprise a weekend, often lasting three or four days. Those weekends are frequently lost to the volunteers as no official rule for time in lieu is in place.

During the fieldwork in Malaysia, 22 dive professionals were interviewed. This included dive instructors, dive masters and dive master trainees from Canada, the UK, Finland, the USA, the Netherlands, Denmark, Sweden and Switzerland. They were aged between 18 and 39, with the majority being between 21 and 28. The majority had a university degree. Often the dive professionals started out as regular backpackers, who took up diving during their trip. They work the tourism seasons between the west coast of Thailand and the east coast of Malaysia, which differ due to the monsoon, or change locations, in the South-East Asian region and beyond, at six-month intervals.

Their living conditions are very similar to the NGO volunteers. They often share rooms above the dive shop or live in simple beach bungalows, both with limited generator-powered electricity. They have no prescribed holiday allocation, and need to take time off to do 'visa runs' every three months.

The dive professionals fund their own stays. The dive instructors work based on commission and so need to sell dive courses to tourists. These commissions cover their basic living costs and costs of transport to their next work location, but do not allow for real savings or luxuries. Dive masters and dive master trainees are entirely self-funded and have often saved up prior to a long backpacking trip. Especially during the peak season the dive shops get very busy and it is not uncommon that the staff are forced to work many days in a row without a break.

The *Caretaker Gazette* confirms the existence of a volunteer market in the hospitality sector. Initially set up for finding employees for the care of houses and properties, more and more advertisements are placed by hotels, guesthouses, bars and restaurants for volunteer managerial staff. These posts are promoted as enabling the caretakers to 'live where others holiday', thus drawing on a similar imagery as can be found in the volunteer tourism project market. Similar to the two other groups in this comparative study, these volunteers get room and board in exchange for their work, in some cases in combination with a small stipend to cover initial travelling costs. Two such volunteer caretakers were interviewed for this research and their answers were similar to the other interviewees.

Findings and discussion

Motivation

Table 3.2 highlights the similarities and differences in the volunteer motivations stated by the interviewed volunteers. The rankings were determined by first of all counting the frequency of the statements. As the interviews were informal and unscripted by nature, some of the volunteers indicated their reasons prior to being asked explicitly. Any repeat statements in the same subject area of the same person were not counted again. A mentioning of 'my main reason', or similar emphasis on one of the motivations, was given additional weighing.

For the volunteers in Nicaragua, the main driver was the enhancement of their curriculum vitae (CV), gaining new skills and field experience, in order to find employment within their desired sectors. Many of them stated that they aspired to a career with an NGO or as an engineer in rural development. One of the volunteers needed the experience as a credit in his university degree, and one was doing it as part of his doctoral research. Work-related motivation was ranked highest amongst the volunteers in tourism and dive professionals as well, with the exception that they were not doing the work in order to enhance their CVs for improved chances in the job market, but to get away from a job or career that had left many of them with a sense of frustration. This same sense of frustration could be noticed amongst the Nicaraguan volunteers, many of whom had previously worked in regular careers. Yet, rather than turning their backs completely on the professions that they were trained to work in (with one exception), they attempted to gain additional skills and experiences to build on their existing training in order to achieve a career change or a change from the private into the NGO sector.

The second most frequently stated reason by the Nicaraguan volunteers for undertaking their long-term commitment was the desire to travel the world and live abroad for an extended period of time, a motivation which ranked on third place with the dive professionals. However, the latter's second motivation, the holiday lifestyle or to extend their backpacking trip, is not dissimilar to the wish to travel. As stated previously, many of the dive professionals entered the profession during a long backpacking trip. As part of the trip they took up diving and in order to continue diving they decided to enrol in professional dive training whilst working for a dive shop. One dive instructor summed it up as 'the best and easiest way to get paid and tour the world'.

Table 3.2 Volunteer motivations

Aspect	VOLUNtourists – those in development project	VolunTOURISTS – those in diving and hospitality
Motivation I	• CV and self-enhancement	• escape the rat race
Motivation II	• travel the world and live abroad for longer	• holiday lifestyle; extend backpacking trip
Motivation III	• don't know what else to do	• travel the world and live abroad for longer
Motivation IV	• do something useful	• don't know what else to do

For the Nicaraguan volunteers on the third place and for the dive professionals on the fourth place ranks possibly the most striking of motivations, the fact that the interviewed young adults did not know what else to do with their life. When talking about this, particularly with the Nicaraguan volunteers, the same sense of frustration was revealed as with the motivator career enhancement. The respondents felt that despite a solid university education, and often several years of work experience in their profession, they found themselves at an impasse. As one volunteer, who was completely disillusioned with her original career and work in the private sector, stated 'I am taking this year to find out what I really want to do with the rest of my life'.

Only as fourth ranked motivator the Nicaraguan volunteers stated that they wanted to 'do something useful'. This altruistic motive was only mentioned by eleven of the sixteen interviewed volunteers, and not one of them gave it as their principal motivator. This is striking, showing clear discrepancy with what we might expect when thinking of volunteers going to less developed countries. In several cases it was even considered a consoling factor. As one volunteer said 'at least this way I am doing something useful', which he felt had not been the case in his previous professional career.

The desire for young adults to travel and explore the world is, in itself, not surprising and has a long history dating back to at least the European 'grand tour' in the eighteenth century and more recently in the hippy overland trail of the 1960s and the more recent flows of young backpacker tourists in South-East Asia and elsewhere (Hampton, 1998; 2003). The high level of frustration and general sentiment of unhappiness with what we might term a 'traditional' professional life in the West, however, is alarming, especially as it concerns well-qualified individuals, who chose not to be part of it.

Overall impacts

One aspect of the voluntourism definition according to Wearing is 'aiding or alleviating the material poverty of some groups in society, the restoration of certain environments, or research into aspects of society or environment' (2002: 240). Table 3.3 offers a brief overview of the positive and negative impacts relating to the VOLUNtourists and volunTOURISTS. The details are now discussed.

Table 3.3 Impacts

Aspect	VOLUNtourists – those in development project	VolunTOURISTS – those in diving and hospitality
Positive impacts	• projects for local development	• unintentional impacts
Negative impacts	• filling jobs locals could fill • cultural tensions between volunteers and hosts	• remove pressure on dive shops to train locals • cultural tensions between dive professionals and locals

The VOLUNtourists in this study, i.e. the NGO volunteers in Nicaragua, worked on projects for rural development. Through their work several communities in the area now possess alternative energy sources and water purification systems.

In case of the volunTOURISTS, i.e. the dive professionals, there is no broad intended positive impact, although there was mention of the international dive instructors being able to help in the training of local dive masters. There are however positive impacts that can be noted as a consequence of actions undertaken by international-voluntour dive professionals. It was noted during the research that in general they tended to be more passionate about environmental issues and were often the drivers behind local environmental action, such as beach or reef clean-ups. The stronger environmental awareness was particularly visible in the question of shark consumption, with many international dive professionals boycotting restaurants serving shark. Another positive impact is the spending of the volunTOURISTS, who like other tourists, often stay in beach bungalows, providing a longer term income to the bungalow operators than regular tourists would. They also eat in the tourist restaurants. However, none of the interviewed international dive professionals frequented the local shops in the communities on a regular basis. Instead they sourced their supplies from the mainland.

Negative impacts were similar in both groups. Even though the organisation claimed it was unable to undertake their work without free labour and despite some paid jobs being created, locals accused the NGO volunteers of stealing work. On a similar note, the availability of international volunteer dive professionals means that dive shop owners have no pressure to train locals. As one Malaysian dive shop owner pointed out 'I'd prefer more local dive masters and local instructors but they are hard to get. Locals can't afford the instructor course.' It seems to be a vicious circle as the businesses claim they cannot find trained local staff, so they hire the volunteers, which in turn makes the training of new staff unnecessary.

Both groups furthermore stated cultural tensions between themselves and the locals, often as a result of what we can term 'typical tourist behaviour', i.e. excessive alcohol consumption, partying and sexual liberty. In both cases, interviewees stated that problems started to arise when local friends joined their parties or imitated their behaviour, which led to a clash with their own societies, notably family and elders. As a Malaysian interviewee described, 'when foreigners come here the locals copy their style. For example, the bar is for foreigners but the local people join them and consume alcohol', which causes particular offence to the traditional Muslim population of the village on the island. Several volunteers in Nicaragua mentioned the isolation several local young women faced after they had been involved with international volunteers. Their liberal behaviour was unacceptable to the predominantly Catholic and conservative society in town. As a consequence, in both locations the international volunteers were blamed for corrupting and influencing the local youths.

The volunteer caretakers also stated impacts similar to the dive professionals. Their positive impacts were raising environmental awareness as they were the key drivers of several local environmental projects together with several expatriate hotel owners. They also implemented a new environmental policy in their own establishments and were recruited to provide training in hospitality to local staff. Negative impacts were conflicts with the local population as they were considered intruders, heavily criticised for doing a job for free, which a local person could have been paid for instead.

The voluntourism component

Within the emerging voluntourism debate, the tourism component of these international volunteers is of significance. We have already seen that the motivation to travel and, for the dive professionals, to lead a holiday lifestyle ranked high amongst the list of motivators. Besides these personal statements, the volunteers are officially designated as 'tourists' in the two countries in question. They obtain tourist visas on arrival, stay the allotted three months, then travel to a neighbouring country, often Thailand and Costa Rica respectively, and then re-enter to be issued with a new three-month tourist visa.

Both groups however feel offended when they are referred to as 'tourists'. The dive professionals consider themselves as backpackers, who in their eyes are different from tourists. They have no problem being referred to as 'backpackers'. The Nicaraguan volunteers, on the other hand, have a strong volunteer identity. They appear to have a similar definition of 'tourists' in their minds as the dive professionals had, that is individuals who sunbathe on the beach all day long, and are very quick to emphasise that when they go abroad they are *travellers*, not tourists. When it was pointed out to them that they actually also hold a tourist visa and travel extensively during their stays, especially for the visa runs in their holidays, they still hold firm that they are there as volunteers and do not want to be considered as being tourists, which confirms similar statements in research by Scheyvens (2002) and Mowforth and Munt (2009).

Despite the individuals' own attitude towards tourism, especially mass tourism behaviours and attitudes, the groups in question still match the image of voluntourists. The tourism component is confirmed as travel is a major motivator and their official status is that of a tourist, and they carry out work as a volunteer during their stay and have an impact on social development or environmental awareness. This was to be expected in the case of the NGO workers, as their main difference to traditional voluntourism participation is the duration of their stay, whilst the undertaken activities are comparable. For the volunteer dive staff and caretakers in the hospitality sector however, this is more of a surprise. Nevertheless, their volunteering also shows comparable impacts on local social and environmental issues. The voluntourism definition, i.e. 'those tourists who, for various reasons, volunteer in an organized way to undertake holidays that may involve the aiding or alleviating the material poverty of some groups in society, the restoration of certain environments, or research into aspects of society or environment' (Wearing, 2002: 240), is fulfilled in both cases.

Conclusions

The results showed that both groups of long-term volunteers that were studied had similar motivations to undertake the work in question, were faced with almost identical basic living conditions, and had similar positive and negative impacts on their surroundings. The main difference could be found in their self-perception with one group firmly negating the use of the term 'voluntourist', and there particularly the tourist component, often associating themselves with the locals and wanting to be accepted as such, whilst the other group clearly saw itself as a group of outsiders and commented on the chosen lifestyle as enabling them to travel permanently.

The analysis of the motivation and the impact of participants in both types of volunteer activity clearly showed that Wearing's (2002: 240) definition of volunteer tourists applied to the two groups. Both the volunteer dive professionals and the volunteer NGO workers stated travel abroad or a holiday lifestyle as key motivators and both have an impact on local communities, be it intentional as in the case of the NGO workers or as a side effect in the case of the dive staff. This is important to note, as it widens the group of volunteer tourist activities not only to longer term stays, but also to a new set of participants: the volunteers working in the tourism industry itself. This is significant, not only because of increased numbers of voluntourists overall, but furthermore because it sheds light onto a new group of worldwide volunteer workers.

Whilst in traditional voluntourism NGO organisers can only carry out the projects with free labour and commercial organisers use at least part of the money they are paid by the participants to fund the projects, the volunTOURISTS in this study occupy actual jobs that would have to be remunerated were it not for their free work. In the case of this research, the dive shops and the hotels with the volunteer caretakers would actually have to employ paid local staff if no volunteer workers were available. The lack of qualified local staff, especially in the dive tourism sector, but also in hospitality, was often put forward by interviewed business owners. Be that as it may, it is clear that as long as these volunTOURISTS are available, there is no real push for employers to train locals. Not only does this mean a lost opportunity for the economic strengthening of a local community, but in real terms also a tax revenue loss for the countries in question, as local staff would receive regular remuneration and as a consequence would be liable for income tax.

The study raises a number of questions that will require further research. Initially, the study needs to be tested on a larger sample of long-term volunteers and should particularly include more volunteers working in private tourism companies, as these have so far not been researched at all. It needs to be explored what management implications the different motivations of the volunteers have and how management can address this in order to optimise both the day-to-day running and long-term sustainability of projects and businesses. The impacts of volunteers on local communities also need to be examined. This is an area that has so far been mostly neglected by researchers in voluntourism. It will be of particular interest to see whether there are any differences between the two groups and between long-term and short-term voluntourists in general. Falling into this aspect the most important question to be raised is the one about the long-term volunteer tourists in the tourism sector itself and their impact on local job creation.

Acknowledgements

The authors are grateful to the Faculty of Social Sciences, University of Kent, for the grant assistance for fieldwork in Nicaragua and the British Council for its PMi2 Research Cooperation Award 2008–10 for the research on dive tourism in Malaysia. The authors would also like to thank all the interview respondents who generously gave up their time. The usual disclaimers apply.

References

Brown, S. and Morrison, A. (2003) Expanding volunteer vacation participation: an exploratory study on the mini-mission concept. *Tourism Recreation Research*, 28: 73–82.

Callanan, M. and Thomas, S. (2005) Volunteer tourism – deconstructing volunteer activities within a dynamic environment. In M. Novelli (ed.) *Niche Tourism Contemporary Issues, Trends and Cases,* Oxford: Butterworth Heinemann, pp. 183–200.

Coghlan, A. (2007) Towards an integrated image-based typology of volunteer tourism organisations. *Journal of Sustainable Tourism,* 15(3): 267–287.

Gilmour, J. and Saunders, D. A. (1995) Earthwatch: an international network in support of research on nature conservation. In D. A. Saunders, J. L. Craig and E. M. Mattiske (eds) *Nature Conservation 4: The Role of Networks,* Beatty and Sons: Chipping Norton, pp. 627–633.

Gunn, C. (1988) *Vacationships: Designing Tourist Regions*, New York: Van Nostrand Reinhold.

Hampton, M. P. (1998) Backpacker tourism and economic development. *Annals of Tourism Research*, 25(3): 639–660.

Hampton, M. P. (2003) Entry points for local tourism in developing countries: evidence from Yogyakarta, Indonesia. *Geografiska Annaler B: Human Geography*, 85(2): 85–101.

McGehee, N. G. and Santos, C. A. (2005) Social change, discourse and volunteer tourism, *Annals of Tourism Research*, 32: 760–779.

Mowforth, M. and Munt, I. (2009) *Tourism and Sustainability. Development, Globalisation and New Tourism in the Third World* (3rd edn), London: Routledge.

Mustonen, P. (2005) Volunteer tourism: postmodern pilgrimage? *Journal of Tourism and Cultural Change,* 3: 160–177.

Scheyvens, R. (2002) Backpacker tourism and Third World development, *Annals of Tourism Research,* 29(1): 144–64.

Stoddart, H. and Rogerson, C. (2004) Volunteer tourism: the case for Habitat for Humanity in South Africa, *Geo Journal*, 60: 311–318.

The Caretaker Gazette, http://www.caretaker.org/ accessed throughout February 2009.

Tomazos, K. and Butler, R. (2008) Volunteer tourism: tourism, serious leisure, altruism or self-enhancement, presented at the CAUTHE Conference, Griffith University, Gold Coast, Australia.

Wearing, S. (2002) Recentering the self in volunteer tourism. In G. M. S Dann (ed.) *The Tourist as a Metaphor of the Social World*, Wallingford: CAB International, pp.237–262.

Wearing, S. (2004) Examining best practise in volunteer tourism. In R. Stebbins and M. Graham (eds) *Volunteering as Leisure, Leisure as Volunteering: An International Assessment*, Wallingford: CAB International, pp.209–244.

Webb, D. (2002) Investigating the structure of visitor experiences in the Little Sandy Desert, Western Australia, *Journal of Ecotourism* 1(2&3): 149–161.

4 Journeys of the self

Volunteer tourists in Nepal

Eugenia Wickens

Introduction

The increasing interest in volunteering and the benefits of participating in voluntary activities and projects in a local community are now well documented (e.g. Wearing, 2002; Simpson, 2004; Wearing *et al.* 2005). There is also a growing body of literature on the value of volunteering to the host community. Often viewed as an essential part of modern life, international volunteering is increasingly recognised for its contribution to the welfare of communities. Volunteers are drawn to projects in third world countries, engaging in activities such as restoration of buildings, cultural preservation or teaching in schools. Nepal provides a good example of this because there is a ballooning desire for English language teachers to help out in schools or hostels. Organisations have sprung up to recruit volunteers who can speak English fluently and although there is a preference for native language speakers, such organisations accept volunteers from all over the world.

Following a brief review of the literature on contemporary volunteer tourism and volunteers' experiences, this paper presents some results from an ethnographic study of volunteers working in social and educational projects within the Kathmandu Valley, Nepal. Fieldwork shows that they are drawn to Nepal for a variety of reasons, but largely for the experience of living 'authentically' in a developing country that is a world away from their own culture. They are attracted to the fact that whereas tourists are confined to the tourist route of hotels, tour guides and the tourist bubble, volunteers get a genuine experience of what it is like to live in and around Kathmandu.

The Kathmandu valley in Nepal is a very popular destination for volunteer tourists. First, there are many NGOs and other voluntary agencies offering the opportunity for volunteers to participate in a community-based project in the valley. Second, Nepal, often described as the 'land of Mount Everest' and the 'land of cultural diversity', is known to draw visitors both because of its natural attractions and its cultural heritage. Nepal undoubtedly is one of the world's greatest trekking and white-water rafting destinations attracting visitors interested in outdoor recreational activities. The Everest region is the main trekking destination east of Kathmandu and the most popular. Moreover, the Kathmandu valley has a plethora of heritage sites, including Newari traditional villages and Hindu and Buddhist holy places. The artistic richness of the valley is reflected in the Unesco World Heritage sites such as the 'Monkey Temple', the Durbar squares of Kathmandu, Patan and Bhaktapur, Bodhnath, the Pashupatinath Temple. Visitors are attracted from all over the world, with most arrivals originating from India, the UK, Germany, Holland, the USA, South Korea and Australia (Shrestha, 2002).

Literature review

Volunteer tourism has grown considerably over the last fifty years and projections indicate that this will continue for the foreseeable future (Wearing, 2002; Halpenny and Caissie, 2003). As a phenomenon contemporary volunteer tourism has been approached from a variety of disciplinary perspectives, including social work, sociology, psychology and anthropology, as well as being the subject of a number of multidisciplinary studies (Smith and Elkin, 1981; Caissie and Halpenny, 2003; Simpson, 2004; Stebbins, 1996; Wearing and Neil, 1999; Weiler and Hall, 1992; McGeheea and Santosa, 2005; Jenner, 1982). While there is a plethora of definitions, 'volunteer tourism' is often conceptualised as a genre of alternative and responsible tourism based on the search for and participation in socio-cultural experiences. A defining characteristic of alternative tourism is that it is small in scale and requires little specialised infrastructure (Wearing, 2002). However, unlike other responsible travellers, volunteers spend considerable amounts of time living and participating in a variety of social, educational and environmental projects. The choice of work undertaken is often morally driven and volunteers' intentions and desires are to help less fortunate people (Stebbins and Graham, 2004). Consistent with these previous studies, this paper conceptualises volunteer tourism as a subtype of alternative/responsible tourism, which involves an international journey and a lengthy period of stay in a culture significantly different from that of the volunteer. The term 'volunteer' carries with it notions of 'aiding or alleviating the material poverty of some groups in society', and 'the restoration of certain environments' (Wearing, 2001: 1). By conceptualising it in this way, this paper attempts to shed some light on the topic of how individual volunteers actually give meaning to their experience of the visited host community.

A review of the literature shows that there is a tendency among researchers to adopt one of two opposing orientations in their analysis of the volunteer experience. One perceives the volunteer tourists as 'colonialists' of the visited host community and volunteer tourism as a form of 'imperialism' (Barkham, 2006; Nash 1989). It is assumed that international volunteering is a new form of metropolitan dominance over 'undeveloped' or 'developing' Third World countries. From this perspective volunteer tourism does not contribute to the long-term development of host communities (Said, 1989). It is seen as representing consumer capitalism at its worst. The other sees volunteers and, in particular, those who receive organised guidance as making significant contributions to communities in a sustainable and constructive manner (Broad, 2003). It is argued that volunteering offers the opportunity to gain life skills by helping people in less developed countries. 'The goals of volunteer tourism are generally to provide sustainable alternative travel that can assist in community development, scientific research or ecological restoration' (Wearing, 2002). These two orientations are best seen as the ends of a spectrum of opinion with a range of views lying between them (e.g. Uriely *et al.*, 2003; Simpson, 2004; Sylvester, 1999; Thomas, 2001; Wagner, 2004).

Furthermore, some researchers have drawn a distinction between 'professional volunteers' and 'gap year volunteers' (Simpson, 2004). The 'professionals' volunteer for approximately 1–2 years and often receive some payments for their work. They tend to be well qualified in their field of expertise and often have substantial professional experience (e.g. UN Volunteers). In contrast, 'gap year volunteers' receive no payment for their work and their placements usually last anywhere between one to six months. They rarely have professional experience or qualifications and pay considerable amounts of money to the agency that organised the placement. This classification scheme has proved useful for the purposes of this study and has guided fieldwork by providing me with a sufficient degree of structure within which to proceed, while at the same time aiding the interpretation of the qualitative data. Due to word limitations the paper focuses the discussion on the motivations and experiences of gap-year volunteers participating in social and educational community projects.

Another feature of the existing literature is that many studies treat the 'volunteer' as an object of analysis rather than as a subject with feelings, experiences, memories and stories to tell (e.g. Woods, 1981; Caissie and Halpenny, 2003; Campbell and Smith, 2006). Despite the quantity of studies, the scholarly study of volunteers and their experiences is impoverished by the absence of sustained and empirical consideration of this phenomenon from the volunteer's point of view.

Study methods

The overall aim of this study, which was undertaken in the summer months of July and August 2008, was to gain an understanding of the experiences of volunteer tourists in Nepal. Fieldwork took the form of semi-structured interviews (between 0.5 and 2.0 in duration) with volunteers (46) and placement staff (12), and volunteer self-completed questionnaires comprising open questions (40, of which 20 were completed by respondents who were not interviewed). Thirty of the interviews were gap-year volunteers and the remainder were the professionals. This paper focuses on only the gap-year volunteers that were interviewed.

Approximately half of the gap-year interviewees stayed in guest houses or rented accommodation and the other half with local families. They were engaging in community-based projects, spending about 6–8 hours undertaking various activities, mainly teaching and looking after children in hostels, hospitals or orphanages. Fees paid by the volunteers apparently contributed to the costs of the programmes that support the visited host communities.

Interviews were conducted at a wide variety of geographically dispersed sites (e.g. hospitals, hostels, hotels) within the Kathmandu Valley. The majority of the interviews were taped with the consent of the participant, and subsequently transcribed. A convenient sample was used for the purposes of this study. This is common in qualitative studies of mobile populations such as volunteers, backpackers, travellers and migrants (Lazaridis and Wickens, 1999). The qualitative data were analysed manually. When looking for differences and similarities in participants' responses to a given question, care was taken that responses to a specific question were analysed with due regard to the context and the conditions which produced them. Significant statements were then selected and coded in terms of theoretical concepts and themes found in literature (such as personal development, authenticity and cultural experiences).

A considerable amount of time was also spent talking to NGOs, charities, etc. Local informants, and in particular, my local field guide, facilitated access to several NGOs and other organisations including UN Volunteers, Advocacy Forum, Voice of Children, Concern Nepal, Child Workers in Nepal, Volunteer Nepal, Volunteer Abroad and Project Abroad. These collaborators not only facilitated the gathering of data, but they also acted as my field tutors helping me to build a deeper understanding of volunteering in Nepal. In this way, they helped me validate, in a general sense, my preliminary findings. This feedback on the 'phenomenological validity' was extremely valuable throughout my research. The validity of my findings is however limited by the sample of people I have studied, by the methods used and by the biography of the researcher. As Giddens (1992a: 49) points out, all 'social research is inevitably flawed ... and all is socially constructed'. While this stricture undoubtedly applies to this study, nonetheless my findings are well founded within the limits discussed and may best be regarded as 'valid in principle or until further notice' (Giddens, 1992a: 49).

Motivations of volunteers

Existing studies on why one really wants to volunteer abroad emphasise several reasons including a desire to 'make a difference', for 'career development' and/or 'personal development', and the pursuit of 'fun and adventure' (Caissie and Halpenny, 2003; Galley and Clifton, 2004; Campbell and Smith 2006). It is also interesting to note that the word 'motive' has been variously interpreted by analysts when discussing motivations of travellers (including volunteers). Psychological studies (e.g. Iso-Ahola, 1982) tend to provide explanations that emphasise those motives which satisfy an individual's needs (and by implication volunteer tourists' needs). Other studies suggest that motivation is socially and psychologically determined in that an individual's home environment plays a key role in influencing his/her reasons for travelling (Dann, 1981). It is interesting to note that past research is based on the works of Dann (1981) who identified and developed the 'push and pull' framework. The consensus amongst many analysts is that the push factors, which are concerned with the socio-psychological status of an individual traveller, and the pull factors found in destinations are essential for understanding what motivates people (including volunteer tourists) to travel (Wearing, 2001; Cassie and Halpenny, 2003).

From the perspective of this research, the push and pull factors have to be inferred from qualitative data both conversational and observational. When asked why they wanted to volunteer in Nepal, one participant (B) reported:

I am interested in development work … and because it is often said that Nepal is the poorest country in the world, and so my help will hopefully be worthwhile. Nepal's geography and beauty also attracted me. I wanted to experience a culture different from my own … One of the unique things about the SPW Nepal programme was that the volunteers are placed with host families … the idea of being in constant contact with Nepalese also appealed to me …

The themes of 'personal development', 'interest in development' and experiencing the Nepali culture are also evident in the following extract from an interview I had with this participant (D):

Nepal … is a destination which is rich in its culture and tradition … a destination which is untouched by modernity … the country is beautiful and I was interested in the culture … The wish to gain a personal experience, a general interest in volunteering and development work … and to help with development in a poor country was my other reason …

Other participants reported similar motivations, for example:

I am interested in development work but to see Nepal. … Nepal specifically because less emphasis is put on its development than other places especially Africa. I had four months free during the summer in between my 2nd and 3rd years at uni … and I have always wanted to visit Nepal and experience the culture and I saw an opportunity to go and teach there, but also help in a hostel when the charity was advertised at uni.

(Participant A)

I've always wanted to travel abroad and learn about another culture, and I thought Nepal would give me the most opportunity to be actively involved in development work and get to experience a new culture.

(Participant F)

I wanted to do a gap year type programme … I chose a development charity based in London first and then had to decide which programme to do … I chose Nepal because I suppose I had a desire to help those less fortunate than myself and try to make a difference, do something useful with my gap year. Nepal also appealed because I wanted to experience a different culture.

(Participant L)

Fieldwork clearly shows that for these participants the push factors for volunteering their labour was 'learning about development work in Nepal' and 'personal development'. This finding is consistent with that of other studies (e.g. Galley and Clifton, 2004). However, 'experiencing the Nepali culture' was the dominant pull factor for the majority of volunteers and as the following discussion shows they wanted to work and live 'authentically' in a developing country that is a world away from their own. This finding challenges the view that all gap-year volunteers are in pursuit of fun and adventure (Caissie and Halpenny, 2003).

Volunteering as a cultural experience

While the information obtained from each participant varied considerably, in that each interview captures a unique story, nonetheless, it is the case that many participants 'go native' in Nepal, by experiencing first-hand the day-to-day life of their hosts. Consequently, volunteer tourists are more likely than other tourist types to learn something about the Nepali way of life, its customs and some of its traditions. The following interview extracts, each from a different respondent, illustrate this:

> Nepal is so rich in its culture and tradition … life for many is extremely hard … but Nepali people are honest, genuine, and so friendly it is a unique place.
>
> (Participant L)

> it is a splendid country with diverse culture and nature … I also had the opportunity to be immersed in the Nepali culture … eating with the children, and breathing their entire life everyday.
>
> (Participant G)

Throughout my fieldwork, I encountered participants who stated that they had made an attempt to learn the Nepali language. Conversing with their hosts, using a few Nepali words certainly facilitates encounters between volunteer tourists and their hosts. The following quotations (each from a different interviewee) illustrate this:

> It is important to learn the language so you can understand the traditional lifestyle, the culture and the Nepali people.
>
> (Participant A)

> You can get more out of your visit by learning Nepali, it is the only means by which you can connect yourself to people.
>
> (Participant C)

> Yes I am understanding more about the Nepali people and their traditions, and their culture … I have learnt a lot more about lots of things in Kathmandu, about relationships … the Nepali view of the world, their traditions and rituals, their Hindu and Buddhist beliefs … the cow is the holy animal and killing a cow brings a jail term …
>
> (Participant A)

When asked how they would describe the overall experience as a volunteer common responses were:

It is a wicked place; Nepal has so much to offer and yes I would recommend a working visit to anyone who is willing to work hard to achieve things there ... anybody can volunteer. I think international volunteering is a great way to contribute to populations outside your own community. It allows the volunteer to expand themselves personally while helping others in need ... you also learn about others cultures ... and this can be a rewarding experience ...

(Participant B)

Immensely rewarding, I loved it ... great opportunity to put something back into the community; it has made me more culturally aware and adaptable possibly more independent as well; it has been enjoyable – though also frustrating as there is so much that needs doing.

(Participant C)

It was the right choice for me ... teaching the hostel children gave me more satisfaction than anything else in my life ... a great experience, and you learn so much more about Nepal and its people by actually immersing yourself in their lives ... Challenging but in a good way. Uncomfortable, but rewarding. One of the steepest learning curves of my life so far. Something I would love to do again.

(Participant F)

Our host family made us feel at home and treated us as family ... they were always pleasant ... I worked for three months and my time spent on teaching at the school was time well spent ... very enjoyable. I know I will return to Nepal again, because Nepal has the nicest people and the culture ... I enjoyed it, the whole experience.

(Participant L)

Culture shock

Although participants reported that you get a genuine experience of what it is like to live in Nepal, many volunteers also experience culture shock and in particular when they first arrive. This is what a volunteer said:

Yes to begin with, but I quickly became used to the streets, buses and shops and traditions. I witnessed cultural differences ... such as women are viewed as inferior to men and, as a result, are expected to adopt subservient attitudes ... we are expected to wear clothes that cover our shoulders and legs whilst working in villages.

(Participant H)

Similarly:

Just when I first arrived, and witnessed all of the sights and sounds after the long journey there ... cows intersperse with the people, the traffic and the dust ... spitting by everyone and everywhere and nose picking again by everyone ...

(Participant B)

Throughout my fieldwork, the theme of culture shock was commonly encountered in conversations with participants. This theme is illustrated in the following extract from a recorded interview with another volunteer:

The traffic is unbelievable ... sitting on the top of the buses is not a safe way to travel but it is often our only option ... it did feel rather strange, coming from a multicultural place like England. I was working in a small village and I did feel like I was a source of curiosity for many of the villagers and didn't feel prepared for the many curious looks I got during my walks around the village ... I recall feeling quite insecure and lonely during my first couple of weeks in Nepal.

(Participant G)

Or again:

Yes – first day I arrived in Kathmandu I wanted to leave. The cars, pollution and noise of dogs barking seemed horrific. There seemed to be no order and complete chaos!

(Participant D)

Although there were numerous other stories about volunteers experiencing culture shock the general consensus amongst participants was that they experience the 'real life' of their host community. These findings give empirical support to the view that volunteers can achieve an authentic experience in less developed countries such as Nepal (Emmons, 2006).

Journeys of self-discovery

Embarking on a journey of self-discovery, all participants agree that they learnt something about other cultures in the purer and simpler lifestyles of their hosts. However, fieldwork also shows that they learn something about 'themselves'. How the experience of volunteering impacted upon the self was a frequent topic in my discussions with these participants:

Perhaps I became a little more reflective, patient and also assertive ... I think experiencing a radically different culture also helped me understand better the constructed and provisional nature of my own identity. I suppose this naturally led me to think that my own self could be improved and worked upon (realising it wasn't fixed in stone) and that I could learn from the best traits of my Nepali friends.

(Participant N)

Or again:

> Being a long way from home and friends while experiencing a different culture has definitely taught me more about myself, learned to work in team with people I did not know well before and got to know them better … Nepal has been a great experience and definitely I would recommend anyone to come here and change their life as mine has changed.
>
> (Participant R)

The following extract from an interview with a participant illustrates a common experience:

> The volunteer programme strengthened my own identity and made me more aware of who I was and where I came from. I certainly felt like that by the end of the programme. By the end of this complete immersion into Nepali village life I could also see much more clearly the constructed nature of identities. I could see how easy, with a few more years volunteering, it would have been to 'become Nepali'! I suppose rather than losing something I would say I gained an insight into what makes me 'me'.
>
> (Participant D)

These findings concerning the impact of volunteering on tourists are consistent with other studies (Wearing, 2001; Ellis, 2007).

Conclusion

This chapter has presented and discussed the motivations and experiences of a number of gap-year volunteer tourists interviewed in Nepal. Fieldwork shows that participants are drawn to Nepal for a number of reasons but largely for the experience of living authentically in a developing country that is a world away from their own. They are attracted to the fact that whereas tourists are confined to the tourist bubble, volunteers get a 'genuine experience' of what it is like to live in the country. A common theme amongst them was that volunteering is a meaningful and self-fulfilling experience that goes beyond the limitations of normal holidays. Fieldwork supports the view that volunteer tourists can achieve an authentic experience and insights into the lives of their hosts. However it should be noted that the findings are represented as 'second order constructs', and as such this study does not make any claim to generality beyond the temporal and geographical location of the study itself. Hence considerable caution is required when attempting to extrapolate from the findings of this study. Hopefully, however, the study will provide a stimulus for further research. It should also be noted that this paper reports preliminary conclusions drawn from the analysis of a subset of all of the data obtained during this field trip. Further reports should be anticipated.

Acknowledgments

I would like to thank my Field Guide, Rabindra Tuladhar, who runs the Nepali side of the NGO Help To Educate (www.help2educate.org). Without Rabindra's help, guidance, support and unflinching pursuit of volunteer gatekeepers, this study would not have been possible.

References

Barkham, P. (2006) Are these the new colonialists? *The Guardian*, Friday 18 August.

Broad, S. (2003) Living the Thai life: a case study of volunteer tourism at the gibbon rehabilitation project. *Tourism Recreation Research*, 28: 63–73.

Caissie, L. and Halpenny, E. A. (2003) Volunteering for nature: motivations for participating in a biodiversity conservation program. *World Leisure Journal*, 45: 38–50.

Campbell, L. and Smith, C. (2006) What makes them pay? Values of volunteer tourists working for sea turtle conservation. *Environmental Management*, 38: 84–98.

Dann, G. M. S. (1981) Tourism motivation and appraisal. *Annals of Tourism Research*, 9(2): 187–219.

Ellis, S. (2007) Voluntourism – Pros, Cons and Possibilities, Energize Inc: empowers and inspires leaders of volunteers worldwide, Philadelphia, http://www.energizeinc.com.

Emmons, K. (2006) Meaningful tourism: educational travel and voluntourism. tourism authority of Thailand. Accessed 1 December 2006 at http://www.tatnews.org/emagazine/2874.asp.

Galley, G. and Clifton, J. (2004) The motivational and demographic characteristics of research ecotourists: operation Wallacea volunteers in south-east Sulawesi, Indonesia. *Journal of Ecotourism*, 3: 69–82.

Giddens, A. (1992a) *The Consequences of Modernity*. Cambridge: Polity Press.

Halpenny, E. and Caissie, L. (2003) Volunteering on nature conservation projects: volunteer experience, attitudes and values. *Tourism Recreation Research*, 28(3): 25–33.

Iso-Ahola, S. (1982) Toward a social psychological theory of tourism motivation: a rejoinder. *Annals of Tourism Research*, 9(2): 256–262.

Jenner, J. R. (1982) Participation, leadership, and the role of volunteerism among selected women volunteers. *Journal of Voluntary Action Research*, 11: 27–38.

Lazaridis, G. and Wickens, E. (1999) 'Us and the Others': the experiences of different ethnic minorities in the Greek cities of Athens and Thessaloniki. *Annals of Tourism Research*, 26(2): 632–655.

McGeheea, N. and Santosa, C. (2005) Social change, discourse and volunteer tourism. *Annals of Tourism Research*, 32(3): 760–779.

Mowforth, M. and Munt, I. (1998) *Tourism and Sustainability: New Tourism in the Third World*. London: Routledge.

Nash, D. (1989) Tourism as a form of Imperialism. In V. Smith (ed.) *Hosts and Guests: The Anthropology of Tourism*, 2nd edn. Oxford: Blackwell.

Said, E. (1989) Representing the colonised: anthropology interlocutors. *Critical Inquire*, 15(2):205–225.

Shrestha, H. P. (2002) *Tourism in Nepal: Marketing Challenges*. New Delhi: Nirala Series.

Simpson, K. (2004) Doing development: the gap year, volunteer tourists, and a popular practice of development. *Journal of International Development,* 16(5): 681–692.

Smith, D and Elkin, F. (1981) Volunteers, voluntary associations and development: an introduction. *International Journal of Comparative Sociology*, 21(3–4): 1–12.

Stebbins, R, and Graham, M. (2004) *Volunteering as Leisure/Leisure as Volunteering: An International Assessment.* Wallingford, Oxon: CAB International.

Stebbins, R. A. (1996) Volunteering: a serious leisure perspective. *Non-Profit and Voluntary Sector Quarterly,* 25(2): 211–224.

Sylvester, C. (1999) Development, studies and postcolonial studies: disparate tales of the Third World. *Third World Quarterly*, 20(4): 703–721.

Thomas, G. (2001) *Human Traffic: Skills, Employers and International Volunteers.* London: Demos.

Uriely, N., Reichel, A. and Ron, A. (2003) Volunteering in tourism: additional thinking. *Tourism Recreation Research*, 28(3): 57–62.

Wagner A. (2004) Redefining citizenship for the 21st century: from the National Welfare State to the UN Global Compact. *International Journal of Social Welfare*, 13(4): 278–286.

Wearing, S. and Neil, J. (1999) *Ecotourism, Impacts, Potentials and Possibilities.* Oxford: Heinemann-Butterworth,

Wearing, S. (2001) *Volunteer Tourism: Experiences that Make a Difference.* Wallingford, Oxon: CABI International.

Wearing, S. (2002) Re-centering the self in volunteer tourism. In G. Dann (ed.) *The Tourist as a Metaphor of the Social World.* Wallingford, Oxon: CABI Publishing.

Wearing, S., McDonald, M. and Ponting, J. (2005) Building a decommodified research paradigm in tourism: the contribution of NGOs. *Journal of Sustainable Tourism,* 13: 424–439.

Weiler, B. and Hall, C. M (eds) (1992) *Special Interest Tourism.* London: Bellhaven Press.

Woods, D. E. (1981) International research on volunteers abroad. *International Journal of Comparative Sociology*, 21(3–4): 196–206.

5 Young non-institutionalised volunteer tourists in Guatemala

Exploring youth and self-development

Christian Schott

Introduction

This chapter extends the research on the development of the self as an aspect of volunteer tourism experiences by presenting findings from a study of young non-institutionalised volunteer tourists in a developing country. The research builds on earlier work by Wearing (2001; 2002) and Wearing and Deane (2003) which examined the impact of the volunteer tourism experience on the volunteer tourist. In his research with Australian Youth Challenge International, volunteer tourists engaged in environmentally focused projects in Costa Rica, Wearing (2001) identified self-development to be one of the outcomes reported by volunteer tourists as a result of the experience. In addition to highlighting the significance of personal development experienced by these volunteers, he also illustrated a few such experiences and classified them into four clusters of self-development. However, to further our understanding of volunteer tourism and volunteer tourists more specifically, there is a need to examine the experiences of volunteer tourists by taking account of contextual factors (Pearce and Coghlan, 2008) and to explore the complexity of volunteer tourist motivation (Söderman and Snead, 2008). It has also been acknowledged that 'volunteer tourism' encompasses many different types of volunteer experiences (Callanan and Thomas, 2005) and that volunteer tourists are not a homogeneous group; different volunteers or groups of volunteers seek experiences according to their personal preoccupations and ambitions (Pearce and Coghlan, 2008).

This chapter then seeks to both widen and deepen our understanding of volunteer tourists by exploring the motivations and experiences relevant to the development of the self in the context of a group of volunteers that has not received much academic attention, young non-institutionalised volunteer tourists. This group of volunteer tourists is different to most previous studies on this topic due to a number of factors: the organisation that the individuals volunteered for is a locally run orphanage in Guatemala that is entirely independent of international sending organisations (Casa Guatemala), the respondents stayed at the orphanage for relatively long periods of time (3–12 months), were aged between 18 and 25, represent a range of nationalities, and all volunteered at the orphanage on their own. As such, this (presumably less common) group of volunteers could be described as the volunteer tourism equivalent of Cohen's non-institutionalised travellers (Cohen, 1972) because they are independent of sending organisations, volunteer on their own and have committed to live and work in a poverty-stricken developing country. Attempting to position this group in Callanan and Thomas's (2005) classification framework according to the 'project' characteristics, the volunteering experience at the orphanage is best described as a deep volunteer tourism project; however, some of the classification criteria are debatable.

Volunteer tourism and personal development

The literature on volunteer tourism has expanded rapidly in the years since Wearing published the first book on volunteer tourism in 2001. The literature now discusses many different facets of volunteer tourism, for example, volunteer tourism as a force for peace (Higgins-Desbiolles, 2003), different stakeholder perspectives on volunteer tourism (Gray and Campbell, 2007; McGehee and Andereck, 2009), volunteer tourism as a sustainable form of cultural tourism (McIntosh and Zahra, 2007), a classification framework of volunteer tourism projects (Callanan and Thomson, 2005), the structure of the volunteer tourism industry (Ellis, 2003; Raymond, 2008), the relationship between volunteer tourism participation and interest in social movements (Gard McGehee, 2002; McGehee and Santos, 2005), highly institutionalised forms of volunteer tourism (Brown and Morrison, 2003), as well as numerous case-studies of different volunteer tourism projects. This overview of the literature is of course not comprehensive, but merely a snapshot, with the most notable omission being the chapters and articles focusing on the volunteer tourist; a topic that needs to be discussed with more than a fleeting mention. Since the early publications on volunteer tourism, the motivation and the volunteer tourists' perspectives on the experience have been of great interest to researchers: in the context of 'mini missions' on structured holidays (Brown, 2005), in the context of the British gap year (Simpson, 2004, 2005; Söderman and Snead, 2008), as a reflective study of volunteer tourism experiences earlier in life (Zahra and McIntosh, 2007), in the form of motivational research of wildlife volunteers (Broad, 2003; Broad and Jenkins, 2008), exploring the motivational differences between community and wildlife volunteers (Lepp, 2008), as conceptually focused discussions about altruism and postmodern theory (Matthews, 2008; Mustonen, 2005, 2007; Uriely *et al.*, 2003), and most extensively by exploring perspectives on the development of the self by positioning research findings in the sociology literature (Wearing, 2001; 2002; Wearing and Deane, 2003; Wearing *et al.*, 2008).

Most of the authors who discuss the volunteer as the focus of their work acknowledge that volunteer tourism experiences present outcomes to the volunteers that can be categorised as personal development or development of the self. As such it is now widely recognised that most forms of volunteer tourism are not just about 'doing good for others' but also about 'doing good for self' (Matthews, 2008: 111). Conversely, while recognising that benefits of a self-directed nature occur, the depth and breadth of outcomes that relate to development of the self have not been fully explored. Wearing, who examined the construct of self in some depth in a variety of single and co-authored publications, has contributed the best insight into this topic to date. As such, his key findings and propositions regarding the concept of self-development in the context of the volunteer tourism experience will be reviewed briefly to contextualise the research presented in this chapter. However, it should be noted that this chapter does not seek to engage in the discussion around the complex construct of 'self'; for an insightful interdisciplinary approach to this debate see Wearing and Wearing (2001).

Wearing (2001) analysed his research with Australian volunteers who had participated in a Costa Rican ecotourism project organised through an Australian sending organisation at two levels; the motivations for volunteering as well as the aspects of the experience that relate to the development of the self. The motivational categories he identified are altruism, travel/adventure, personal growth, cultural exchange/learning, professional development, the YCI (Youth Challenge International) programme and right time/right place. Apparent from this list is that the motivations for these individuals to volunteer abroad are wide-ranging and encompass motives that relate to others as well as to the self. Wearing then narrows his focus to the aspects of the volunteer tourism experience that relate to the development of the self and formulates four clusters: personal awareness and learning, interpersonal awareness and learning, confidence and self-contentment. However, whether these interesting findings are relevant to other groups of volunteer tourists, and whether the self-development aspects observed represent motivations or serendipitous by-products of the volunteer tourism experience is unclear.

The specific aims of this research then are to examine the self-development motivations and experiences of a group of young solo volunteer tourists at an orphanage in the Guatemalan rainforest (Casa Guatemala), and to explore whether a self-development process is actively sought or merely a by-product of the experience. To deepen our understanding of self-development, this chapter also explores the diversity and complexity of self-development motivation and experiences, and as such highlights what specifically it is that is developed or sought to be developed in relation to the volunteer self. This discussion will be contextualised in two ways. First, by exploring which stage or 'space' in life the volunteers consider themselves to be in, as a measure of attaining whether the individuals are indeed 'youth', not just as classified by age but also in terms of the life stage 'youth'. Second, by establishing how the volunteers frame the 'trip', in other words what the trip represents or means to the individual.

For the purpose of this chapter, which does not seek to engage in a detailed analysis of discipline-defined terminology, the terms self-development and personal development will be used interchangeably referring to the development of the individual as opposed to the development of 'other'. It should also be noted that the term 'trip' will be used to refer to the entire volunteer tourism experience, including the time at the orphanage as well as travel before, during or after volunteering at Casa Guatemala.

Methodology

To gain an understanding of the deeper-lying thought processes and motivations behind the decision to volunteer and the experience itself, semi-structured in-depth interviews were chosen as the most appropriate research method. Interviewees were recruited from a group of roughly 45 volunteers living at Casa Guatemala at the time of the research, according to their availability to participate in the project (depending on work commitments and leave arrangements) and their English language ability. The interviews were conducted in an informal face-to-face setting at the orphanage. Prior to commencing the interviews the volunteers were given information about the nature of the research and advised that the project had been reviewed and approved by a university ethics committee. A total of nine in-depth interviews were conducted with volunteers from six different countries (Belgium, Canada, Germany, Israel, UK, USA), with each interview lasting between one and two hours.

The nine interviewed volunteer tourists had all arrived at the orphanage on their own with an expectation of staying for periods ranging from three months to a full year. At the time of the interview only minor variations from the originally anticipated length of stay were mentioned. All nine were intending to travel around Central America either before, during or after their time at the orphanage. The age profile positions the respondents in an age category that can best be described as youth because they were aged 18 (1), 20 (2), 22 (4), 23 (1) and 25 (1) at the time of the interviews. The respondents' background in terms of nationality was Belgian (1), Canadian (1), German (2), Israeli (1), US American (3), and British (1). Six of the respondents were female and three male.

Findings

All nine respondents had very different 'stories', both in terms of their backgrounds as well as reasons for volunteering at the orphanage. For example the respondents' culture in addition to several other contextual factors appeared to have some bearing on the findings (as suggested by Pearce and Coghlan 2008); however, to protect the respondents' identity the findings will be presented without attributing quotes to particular individuals or identifying their distinctive background. Rather, this section will build an understanding of commonalities and differences between responses and examine and illustrate the range of aspects discussed in the context of development of the individual. Initially, the chapter will present 'how' and 'where' the volunteers view the self in the context of their life journey, which will be followed by exploring how the nine individuals frame the trip. The examination of the respondents' narratives in the context of self-development will then be presented. This discussion will be supported by applying the four themes (personal awareness and learning, interpersonal awareness and learning, confidence, self-contentment) developed by Wearing (2001) to structure the analysis and assess whether this group of volunteer tourists reports similar experiences.

Volunteers' life stage

Asked to discuss 'where' the individuals would position themselves in terms of their life journey from birth to death, in other words which life stage they considered themselves to be in, and how they would explain and characterise this stage, all nine volunteers appeared to be in a transitionary stage in life. The manners in which the volunteers described the life stage varied in terminology, including quarter-life crisis, time to live, finding identity, a time 'pre responsibilities', a time for independence, a time for liberation, etc. However, in connotation all the descriptions are characteristic of 'youth', or late adolescence as a time between childhood and adulthood (Roberts, 1983: 190), as 'youth is seen as a stage in the life course where individuals have the freedom to find out about the world and themselves as part of a transition into full adulthood of responsibilities and making commitments to others' (in Desforges, 1998: 190).

The volunteers' narratives also had strong associations with Erikson's (1968) construct of psychosocial moratorium; a delay of adult commitments to 'experiment' and explore, which further supports the notion that the nine individuals presented in this chapter are young volunteers in both age and life stage.

Volunteers' framing of the trip

The interviewees were also asked what the volunteer tourism trip meant to them in terms of its significance and role in their lives. This question sought to draw out broader framing themes rather than detailed multi-motivational constructs. Apart from the clearly stated desire to help those that need help the most and to 'give back', a common theme mentioned was 'challenge', in that the experience represented an opportunity to challenge the self. Interestingly, respondents with less experience of travel abroad saw the trip as an opportunity to learn about the individual's response to the challenge per se, while those with previous overseas experience framed the trip in the context of a desire to seek out and 'conquer' new challenges in a continuous quest to 'see what I am made of' and refine character traits and skills. Aspects related to the notion of challenge were also mentioned, such as the trip representing a catalyst in the desire to build greater self-confidence and self-reliance. However, other profound ways of framing the experience were also noted, such as the search to understand one's identity better by returning to the country and place from where the volunteer had been adopted. Also, a space for reflection, orientation and searching of (new) perspectives on life away from the home environment was mentioned by some volunteers. Evidence of the trip representing an opportunity to obtain cultural capital (Bourdieu, 1984) by collecting 'good stories' to take home was also noted. Indeed, the desire to be away from their home environment and many of the things that it represented was a key theme underlying the ways in which the respondents framed the trip in that the trip symbolised unchartered and unprejudiced space. Not only did this take on the form of a desire to 'escape' significant others (referring to people with influence on an individual's self-evaluation) and the familiar support networks available at home, but also to escape political and societal factors in search of something 'to really believe in'. This brief insight into how the volunteers framed the trip provides a clear indication that not only the development of others but also the development of the self were indeed reasons for the trip from the outset. Further, the framing of the trip is consistent with the life stage discussed, in that in personal terms the trip appears to be an extension or an expression of this transitional life stage.

Development of the self

This part of the interview focused on the volunteer's narratives of self-development and seeks to shed light on the range of aspects that the individuals were developing as a result of the volunteer tourism experience. It needs to be noted at the outset of this section that altruism was clearly a dominant reason for the respondents to volunteer at the orphanage. While this reason also has relevance to the discussion about personal development in that altruism is regarded as a personal reward in the context of personal enrichment (Stebbins, 1996), the analysis will focus more specifically on the concept of development of the self. In doing this, the chapter does not seek to examine the strengths and weaknesses of Wearing's (2001) classification of clusters of self-development but will instead use the framework as a means of structuring and contextualising the findings section.

Personal awareness and learning

Wearing (2001) describes this theme to include personal beliefs, values, abilities, limitations and aspects of professional development. The interviews revealed the majority of points discussed in the context of self-development to fall into this theme. The developmental processes discussed encompassed different levels of significance to the volunteers' sense of self, including improvement of professionally focused skills as well as very profound processes that are central to the individual's core identity.

Fostering a language skill was mentioned as one of the benefits volunteers actively sought from the experience. The language skill was seen as something that could be beneficial to the individual's employability afterwards. However, numerous aspects that are more central to the individuals' sense of self were also mentioned and will be discussed now. These aspects included the desire to both test one's beliefs and assumptions in the context of the reality of humanity and to find and embrace an activity that the individual can '100 per cent believe in' and thus fully support.

Foster language skill

Well I really wanted to use my Spanish ... Improve my Spanish

Test one's beliefs

> Maybe this is also me getting my ass kicked period, where I go out with the ideals that are bred by (university education), and get them smashed to pieces a little bit. All given breath and give them air. Kind of the real world discovery period.

Find something to believe in

> I want to work in something I do believe in. Give myself completely to that … My (previous) period, with myself was the hardest time that I had, because I was doing something I wasn't 100% sure of. I had political and moral problems with it. I wasn't as happy as a person as I am now. I believe 100% in what I'm doing now … I believe more in my goals, so I appreciate more what I'm doing now and where I am, than before.

Being more tolerant and understanding of others and the self were also mentioned. The desire to be more supportive of the self in achieving the developmental goals set in order to become the individual one wants to be were discussed as something for which the experience was intended from the outset. The other perspective offered on developing greater tolerance, framed in the context of others and self, was also valued by the individual but identified as an outcome that was not consciously sought from the trip.

Give self more time

> One thing I wanted to do, is give myself more time for that (personal) growth and understanding, which for me meant more time to sit and more time to read. In doing it, is what I set out to do and about two hours a day of reading, which I love, and while I'm reading all I'm doing is considering new ideas.

More understanding and tolerant of self and others

> Well I'd never been a good stress taker, I mean sometimes I'm still not but I think now I've learned that ok the world isn't going to end if things don't go this way, or if I get stressed out about this, its not going to help me at all, I mean I wasn't good with that at all but now I've grown from that … I'll be a lot more open to a lot of ideas, the cultural aspect has opened my eyes as how ignorant some of the people are.

The notion of exploring one's self in terms of identifying strengths and weaknesses was discussed by several volunteers. There was a clear sense of wanting to explore 'what kind of person I am'. These aspects were generally mentioned as driving forces for wanting to take the volunteer tourism trip, as evidenced in the volunteer's framing of the trip. Those that had less experience of travelling and volunteering away from their home environment were interested in self-development by learning about their strengths and weaknesses, whereas those volunteers with more travel and volunteering experience away from home focused this process on exploring the limits of what they were capable of; in the words of one of the respondents 'a competition against myself'. The trip as an opportunity to find one's limits was an important motivation for at least two volunteers and was strongly incorporated into the travel component of the experience. They both felt that solo travel was essential to shed one's fears and insecurities and truly put the self to the test, which relates strongly not only to the framing of the trip but also the volunteers' life stage. One of the volunteers for example intended to walk back to the US from the Guatemalan rainforest alone; a distance of about 1200km, for which the individual allowed 6–8 weeks. The other intended to hitchhike from Guatemala to the East Coast of the US before returning to Europe, by relying exclusively on hitchhiking for transport and couch-surfing arrangements for accommodation.

Identify character strengths and weaknesses

To realise what I am capable of, now, because I don't know what that is. I don't know what I am capable of ... before time runs out and I get involved in something that I don't want to get involved in, and that is not worth my time. So I really want to search out what kind of person I am. What kind of things I am capable of now.

I think I have just figured out a totally different part of myself, a totally different part ... knew that I was really really really challenged, but that's what I wanted, that I was challenged.

Explore depth of personal abilities and identify limits

I know my own limits in a lot of respects. And I know my own social limits ... at least I'm starting to understand my own limits. I mean we never know the full potential of ourselves, what we're capable of and what we are not capable of. Where, in the situations that I have been put in, I have made it a point trying to understand what I'm doing, and why I am reacting in certain ways ... I'm here, I'm getting experience, I am seeing what I am made of, and seeing what I can and what I can't do. And seeing what effect I can have on an organisation and I am gaining real-world experience as they call it ... I'm competing against myself, and it's a situation where I feel that I can test my potential.

I wanted to find out if I can bear loneliness that much, I knew that I can bear it before to some extent. Like I did not have to be with people to be happy, but I did not know whether I could bear to travel alone all throughout Mexico and the United States.

The story of one volunteer was particularly profound in terms of the relevance of the experience to the individual's sense of self and identity. The person was born in Guatemala and lived in the orphanage before being adopted and growing up in the United States, 'it was the sense of always knowing that made me feel that I needed to go and search'. The person's journey to the orphanage had clearly self-directed motivations in addition to strong altruistic motives. As illustrated in the quote below, the young volunteer was exposed to three highly emotional experiences that had significant implications for the self, understanding identity as being of Guatemalan birth parents, the reality for children of living in an orphanage, and what it 'means' to adopt a child.

Greater knowledge of roots and identity

> I knew; it'd be emotional for me to see the kids and see Casa and just come back to a country I'd never been to but I knew I was from, 'cause I didn't know anything about Guatemala really, that there was a civil war, that I was from there and that was pretty much it. But coming back I've learned a great deal more ... I know I'd adopt a child, I've always known that even when from when I was, I don't know how old, but I've always thought that that's such a generous thing for a person to do.

Interpersonal awareness and learning

This theme 'relates to an awareness of other people and, through this, an appreciation of effective communication' (Wearing, 2001: 129). As previously identified by Wearing (2001) some of the developmental motives relate to more than one of the four themes. Such a case is presented in the quote below which illustrates a concern to understand how people perceive and subsequently respond to the individual coupled with a desire to be more aware of interpersonal dynamics and understand whether the individual 'treats people the right way'. Characteristic of the youth stage in life, a sense of 'benchmarking' the self is also expressed.

Improve interpersonal skills

> It interests me what other people think of me, so that I can reflect about it. Is it how they see me, or is it how I am? or why I don't like this person? ... I want to find out if I am treating people the right way, if I handle people the right way and here I get to know so many people. For example I now know that I have been treating some people wrong before I came here ... I want to find out about what is the difference between me and other people and what similarities there are.

A deep curiosity about different attitudes to life and the fundamentals of happiness were also mentioned as a motivation to seek out this experience. Understanding how other people live and whether the individual would and could harmonize with that life-style was also discussed as an aspect that had implications for the development of the self.

Experience other perspectives on 'life'

> I was curious about the Lonely Planet Guide that said that people here have one eighth of what we have and usually have to spend almost all of the money they earn to feed the family and they are still happy even though they have to struggle for being alive every day. That made me curious because I knew before that we have way too much and are used to having way too much because we have grown up with a lot in our childhood. Like I know a lot of friends that had a lot in their childhood and the only thing they do in the rest of their life is to pursue that kind of status … and I knew that I wanted to know something completely different, I wanted to know why the people here are so happy even though they don't have anything.

Self-contentment

This theme is formulated by Wearing (2001) to include a firmer belief in one's self, abilities and skills. A number of the volunteers talked about the experience as helping them to achieve an attitude change, a change to how they feel and view themselves. A greater sense of accepting who the individual is and coming to terms with strengths and weaknesses was reported by one respondent, which led them to be less concerned about the opinions of significant others. Another volunteer talked about a sense of unhappiness in life and a perceived need to go on a volunteer tourism trip to fulfill a dream and overcome this sense of unhappiness. The volunteer experience was also seen as an opportunity to search for a way of applying the self in a manner which would provide a strong sense of success. To the background of a long-standing sense of failure in an academic setting one of the volunteers sought out and embraced the experience as an opportunity to apply the skills that were not perceived as valued beforehand and to thus gain a strong sense of personal success. The applied nature of the tasks that the volunteer performed at the orphanage provided opportunities to generate results that were valued crucially by the children but also by the individual, which provided a long-sought sense of success and self-worth: 'feel like a successful person'.

Less vulnerability to other people's views of self

> I think less of what other people think. I am less occupied with other people's opinions of myself … just to not be embarrassed for anything.

Regaining sense of happiness

> I just wanted to be happy again, and at the end I just wanted to feel like I learned about a place.

Searching for a sense of success

> I really wanted to succeed at the school (tertiary level), but I couldn't, and I stuck at it for a little too long. And it really hurt. And then what really pushed me to this trip was this other school that I tried to go to ... I really enjoy really working hard (at the orphanage) and knowing that I did a really good job at that. I have felt like such a failure in the past, and I can now take life and use it however I want and know that I can be successful in some things.

Confidence

In Wearing's (2001) framework this theme includes a person's sense of confidence in view of the individual's self-assessment as well as in terms of how other people perceive the individual. As discussed, there is a strong interrelatedness underlying the themes of confidence and self-contentment. The following quotes show this interrelatedness as confidence often has a strong bearing on contentment, and vice versa. Directly related to the framing of the trip as a challenge, several volunteers commented that they sought to develop their level of confidence and were looking to become more at ease in terms of relying on themselves in the face of challenging situations. The notion of shedding support networks, or control networks in the eyes of some, was an important component of this process, and the youth life stage. This search for developing greater confidence in the self was linked to a quest for self-reliance and establishing a visible sense of independence. Illustrated by the quotes below, confidence and the related concepts were discussed in a number of different contexts. For example, one person was concerned about a perceived lack of self-confidence that was considered a limitation in view of the individual's chosen medical career. For this volunteer the trip, in its broadest sense, was framed as an opportunity to challenge a perceived lack of confidence and thus affirm confidence in the self by meeting the challenges of travelling and working in an orphanage without a support network. Another volunteer talked about a different sense of confidence: self-empowerment. For this respondent the experience represented a desired process of change that was perceived as very difficult to achieve. Hence, accomplishing the change and shedding the dreaded 'routine' provided the individual with a strong sense of self-control.

Build greater confidence

> I think I always did not have very much self-confidence ... and in my profession (medical) you need to be sure, you need to take decisions right now, sometimes in an acute situation. And if you're not sure, you go maybe this may be that, but you have to be sure ... it is something that I need to be better at, I need to get better at. I think it will be until I am dead, but I think it is better now than I was before.

Personal empowerment

> Like, it's like feeling like if I want to make a change it's my decision to make a change. Now if I get bogged down into any kind of routine I'll know that I can make a change, and I'll stop the routine whereas before I suppose I didn't feel like I had the power to do that.

Another aspect of self-development discussed was fostering a sense of independence. Independence was talked about in terms of a desire to be both independent from parents as well as viewed and treated as an independent individual by them. Interestingly, but not surprisingly, a volunteer specifically discussed that independence and a desire to develop greater self-confidence were considerations when arranging the volunteer tourism experience, which motivated the individual to seek a less institutionalised type of volunteer tourism experience.

Establish personal independence from significant others

> Earlier this year I got more and more eager to get away mostly because of the fact that I was living at home ... it wasn't really working very well, me working and me living at home. My parents couldn't, it's strange, because they couldn't see me as completely independent but I wanted to be completely independent especially after living away from them during the summer. I came back and it was all slightly odd.

Develop independence by avoiding volunteer tourism institutionalisation

> In the UK there is a big industry of gap year projects where you pay a lot of money and they sort out everything for you, like flight, accommodation and a project and you go in a group of people and I really didn't like that idea. First of all 'cause they rip you off and also you would spend all your time with a group of people from your country and you're not as independent, you've got everything laid out for you.

In seeking to specifically draw out aspects of self-development in the thought processes and experiences of these young non-institutionalised volunteer tourists, the research identified a number of other aspects that appeared insufficiently explored if subsumed in the four themes discussed. These aspects were also clearly concerned with the self, but had a stronger focus on the future development of the individual. These points can best be summarised under the theme 'transitional and directional development needs'.

Transitional and directional development needs

The points raised in this context are strongly related to the individual's stage in life as a period to explore, experiment and dream. They also resonate strongly with the individuals' framing of the trip and can thus be considered motivations for the trip rather than unexpected by-products of the experience. Several volunteers talked about the experience as a catalyst for gaining a stronger sense of clarity about 'what's next' in their life journey. This was discussed in the form of seeking clarity about the next phase in life, as well as a sense of readiness for the next life stage. The experience was discussed as important for some volunteers to reflect upon who they are, where they are in life and where they want to be. The removal of the self from the home environment and the aspects of personal awareness discussed earlier were crucial in these discussions. However, a story was also observed where a volunteer had a reasonably clear sense of what they wanted in terms of a place to live and a professional path; however, the person felt unhappy in starting on this path due to unfulfilled dreams leaving a void in their sense of self – 'quarter-life crisis … you live in your dream stage'. The volunteer tourism experience at the orphanage in Guatemala represented this dream and thus the person felt fulfilled and 'ready' to start on the next stage in life.

Gain clarity on next phase in life

that's one of the reasons why I came here, to find out what I'm going to do. Because I thought, maybe I will find it when I am somewhere else, you know … I have drawn such a conclusion on life here, and that's really cool, I kind of figured out what I want to do and what my dreams are, you know.

No, a lot of my friends my really close friends they've all graduated and all got jobs now and a lot of them I've seen, they go into their first job and they're like, 'I hate this, I absolutely hate my first job, it's the worst job I've ever had', and in a lot of aspects that's very intimidating for me, I mean I'm jumping into that field and I want to make sure that I go into a job that I like, it's kinda like learning from their mistakes that they've made, cause they've jumped into the first job they got an offer for and I want to make sure I'm going into a field that I'll appreciate and like.

Gaining sense of readiness for next stage in life

I can say definitely, it's the pre-step to the rest of my life. Because not only what I'm going to study, but also why I am studying it. Also what I'm going to do with it afterwards.

I was really sad and then I realised that it was because of all these things that I wanted to do and so I wasn't happy because I wasn't doing them. But then I was making it like I don't like my job. And then I was like, do I really not like my job? And then I would realise I like those things, I just need to not do them right then.

Aspects directly related to the development of the self in the context of pursuing or committing to a professional life path were also discussed, in the context of work with young children and becoming professionally involved in NGO work. For one volunteer the experience represented the final test to confirm whether a passion for working with children in need would survive the physical and mental demands of living and working in an orphanage; this test had critical implications for confirming or rejecting the respondent's tentatively chosen professional path. The other example had similarly critical implications for the volunteer, but in the context of NGO work more broadly. Both cases illustrate the importance that such an experience can have for the volunteer's professional development, undoubtedly in terms of the experience and skills gained, but crucially also in terms of critical personal decisions with long-term implications for the individual.

Confirmation for choice of work with children

> I was deciding do I really want to go into social work after, or do I want to go into a completely different field of work? I mean this is the test, can I really do this, do I have the will power? do I have the strength? Can I physically work with kids who I know may or may not have a future in another family.

Confirmation for choice of work with NGOs

> My dream was actually to have my own NGO. So part of coming here was also to see how it works.

Discussion and conclusions

This chapter has illustrated the profoundness and complexity of the self-development experiences and motivations of a group of young volunteer tourists. By interviewing volunteer tourists who have largely been neglected by studies of personal development, this research has contributed both breadth, in examining young non-institutionalised volunteer tourists, and depth, in exploring the diversity of developmental processes sought and experienced, to the topic of development of the volunteer tourists' self. Framing the volunteer trip in the context of the individual's life stage added an additional layer of understanding (Pearce and Coghlan, 2008) to this emerging field, as it assisted in positioning the desires, needs and experiences in that 'ambiguous status betwixt and between, neither children nor full adults' (Roberts, 1983: 35). The relevance of the youth life stage as the backdrop to the motivations and experiences of these volunteer tourists may well explain why these volunteer tourists have chosen a less institutionalised form of volunteering and why they choose to face the experience on their own. The notion of challenge, personal development and life orientation are common themes in the findings presented and of crucial significance to young people in accomplishing the transition from youth to full adulthood. As the self is not static (Wilson in Wearing, 2001) a continual process of change underlies the human condition and this process is of utmost importance to young people, of whom some are active seekers of personal development and self-change and others are more passive recipients. A respondent who was very active and conscious in seeking personal development considered a sense of security as a significant inhibitor to the development of the self: 'If there's one fear I have it is of security. I don't like to feel security because as soon as I feel security, I feel like I have stopped growing in certain respects'. A similar sentiment was expressed by another volunteer who emphasised the benefit of travel and volunteering abroad in a more direct manner: 'when I'm in my home I like rot and don't do anything, but when I'm up against so many new things, cultures, people I'm just constantly realising that oh that's about yourself now that's pretty cool'.

This highlights the significance and important role of volunteer tourism to support and enhance some of the communities, species and environments that are most vulnerable and neglected while at the same time providing crucial opportunities for individuals and particularly young people to develop the self and explore their aspirations and direction for the remainder of their life journey, as 'seeking out the new and unfamiliar and going beyond our daily concept of self is an essential step in the development of "self"' (Wearing, 2001: 9). Both the exposure to the unfamiliar or unknown and the shedding of the safety networks and moulding expectations of significant others are commonly observed factors in this process. When considering that the experience of 'youth' is changing as a result of continuous broader transformation in modern society (Furlong and Cartmel, 1997), new avenues of research on both self-development and youth travel are evident.

The profoundness of the developmental processes that are taking place is further support for the wide-ranging benefits of volunteer tourism. While young tourists do not always enjoy positive press coverage and public opinion (Schott, 2004) leisure and tourism are critical to the young person's personal development and sense of identity. As such, the young volunteer who went to Casa Guatemala to 'give back' and explore the person's roots to thus gain a more comprehensive understanding of the self, presented a very profound case of personal development with no hedonistic undertones. Indeed it would be interesting to examine this motivation for volunteering in greater depth; do many volunteers in overseas orphanages have a shared early childhood history with the orphanage? Similarly, what about other types of 'roots volunteer tourism' where a shared sense of identity with the project or cause plays an important role, such as in the study by Ari *et al.,* (2003).

Also noteworthy in the context of this research are the transitional and directional development needs discussed by the volunteers. While this type of non-institutionalised volunteer tourism experience would presumably be classified as project-based leisure in Stebbins' (2005) theorisation of leisure as casual, serious or project-based, due to the absence of a leisure career trajectory, there is evidence that this initially project-based leisure experience is converted into both a professional pursuit by some and into serious leisure by others.

Finally, the aim to explore to what extent the developmental processes discussed are a motivation for the trip rather than an unexpected by-product of the experience has generated significant evidence that volunteers actively seek to develop the self as part of the volunteer tourism experience. The manner in which the volunteers framed the trip provided insight into the minds of the volunteers and thus the aspects they sought from the experience. It has to be noted that some respondents were very conscious and reflective of their self-development needs and intentions, while other volunteers conceptualised their self-development motivations in different contexts, or 'wrapped them up differently' so to speak, or commented that motivations focusing on the self were only becoming conscious to them while in the midst of the experience, but were subconscious at an earlier stage. There are undoubtedly also a number of self-development processes experienced by the volunteers that were not intended outcomes from the experience, neither consciously nor subconsciously, as well as aspects of self-development that were not discussed. While research on personal change and self-development in the context of various facets of tourism has been conducted since the 1970s, more empirical work is needed to explore these concepts in the many different contexts of tourism; to examine the significance of tourism and particularly tourism by young people to personal welfare, social structures and peace.

References

Ari, L. L., Mansfeld, Y. and Mittelberg, D. (2003) Globalization and the Role of Educational Travel to Israel in the Ethnification of American Jews. *Tourism Recreation Research*, 28(3): 15–24.

Bourdieu, P. (1984) *Distinction: A Social Critique of the Judgement of Taste*. Cambridge: Harvard University Press.

Broad, S. (2003) Living the Thai Life – A Case Study of Volunteer Tourism at the Gibbon Rehabilitation Project, Thailand. *Tourism Recreation Research*, 28(3): 63–72.

Broad, S. and Jenkins, J. (2008) Gibbons in Their Midst? Conservation Volunteers' Motivations at the Gibbon Rehabilitation Project, Phuket, Thailand. In K. D. Lyons and S. Wearing (eds) *Journeys of Discovery in Volunteer Tourism: International Case Study Perspectives*, Wallingford: CABI, pp. 72–85.

Brown, S. (2005) Travelling with a Purpose: Understanding the Motives and Benefits of Volunteer Vacationers. *Current Issues in Tourism*, 8(6): 479–496.

Brown, S. and Morrison, A. M. (2003) Expanding Volunteer Vacation Participation: An Exploratory Study on the Mini-Mission Concept. *Tourism Recreation Research*, 28(3): 73–82.

Callanan, M. and Thomas, S. (2005) Volunteer Tourism: Deconstructing Volunteer Activities within a Dynamic Environment. In M. Novelli (ed.) *Niche Tourism: Contemporary Issue, Trends, and Cases,* Oxford: Elsevier Butterworth-Heinemann, pp. 183–200.

Cohen, E. (1972) Toward a Sociology of International Tourism. *Social Research* 39: 164–182.

Desforges, L. (1998) 'Checking out the Planet': Global Representations/Local Identities and Youth Travel. In T. Skelton and G. Valentine (eds) *Cool Places: Geographies of Youth Culture*, London: Routledge, pp. 175–192.

Ellis, C. (2003) Participatory Environmental Research in Tourism: A Global View. *Tourism Recreation Research*, 28(3): 45–55.

Erikson, E. H. (1968) *Identity: Youth and Crisis*, New York: W. W. Norton and Company.

Furlong, A. and Cartmel, F. (1997) *Young People and Social Change: Individualization and Risk in Late Modernity*, Buckingham: Open University Press.

Gard McGehee, N. (2002) Alternative Tourism and Social Movements. *Annals of Tourism Research*, 29(1): 124–143.

Gray, N. J. and Campbell, L. M. (2007) A Decommodified Experience? Exploring Aesthetic, Economic and Ethical Values for Volunteer Ecotourism in Costa Rica. *Journal of Sustainable Tourism*, 15(5): 463–482.

Higgins-Desbiolles, F. (2003) Reconciliation Tourism: Tourism Healing Divided Societies. *Tourism Recreation Research*, 28(3): 407–418.

Lepp, A. (2008) Discovering Self and Discovering Others through the Taita Discovery Centre Volunteer Tourism Programme, Kenya. In K. D. Lyons and S. Wearing (eds) *Journeys of Discovery in Volunteer Tourism: International Case Study Perspectives*, Wallingford: CABI, pp. 86–100.

Matthews, A. (2008) Negotiating Selves: Exploring the Impact of Local-Global Interactions on Young Volunteer Travellers. In K. D. Lyons and S. Wearing (eds) *Journeys of Discovery in Volunteer Tourism: International Case Study Perspectives*, Wallingford: CABI, pp. 101–117.

McGehee, N. G. and Andereck, K. (2009) Volunteer Tourism and the Voluntoured: The Case of Tijuana, Mexico. *Journal of Sustainable Tourism*, 17(1): 39–51.

McGehee, N. G. and Santos, C. A. (2005) Social Change, Discourse and Volunteer Tourism. *Annals of Tourism Research*, 32(3): 760–779.

McIntosh, A. J. and Zahra, A. (2007). A Cultural Encounter through Volunteer Tourism: Towards the Ideals of Sustainable Tourism? *Journal of Sustainable Tourism,* 15(5): 541–556.

Mustonen, P. (2005) Volunteer Tourism: Postmodern Pilgrimage? *Journal of Tourism and Cultural Change*, 3(3): 160–177.

Mustonen, P. (2007). Volunteer Tourism – Altruism or Mere Tourism? *Anatolia*, 18(1): 97–115.

Pearce, P. L. and Coghlan, A. (2008) The Dynamics behind Volunteer Tourism. In K. D. Lyons and S. Wearing (eds) *Journeys of Discovery in Volunteer Tourism: International Case Study Perspectives.* Wallingford: CABI, pp. 130–143.

Raymond, E. (2008) 'Make a Difference!': The Role of Sending Organizations in Volunteer Tourism. In K. D. Lyons and S. Wearing (eds) *Journeys of Discovery in Volunteer Tourism*, Wallingford: CABI, pp. 48–60.

Roberts, K. (1983) *Youth and Leisure*, London: Allen and Unwin.

Schott, C. (2004) Young holidaymakers: solely faithful to hedonism? In K.A. Smith and C. Schott (eds) *Proceedings of the New Zealand Tourism and Hospitality Research Conference 2004*, Wellington: Victoria University of Wellington, pp. 364–376.

Simpson, K. (2004) Doing Development: The Gap Year, Volunteer-Tourists and a Popular Practice of Development. *Journal of International Development*, 16(5): 681–692.

Simpson, K. (2005) Dropping Out or Signing Up? The Professionalisation of Youth Travel. *Antipode,* 37(3): 447–469.

Söderman, N. and Snead, S. (2008) Opening the Gap: The Motivation of Gap Year Travellers to Volunteer in Latin America. In K. D. Lyons and S. Wearing (eds.) *Journeys of Discovery in Volunteer Tourism: International Case Study Perspectives.* Wallingford: CABI, pp. 118–129.

Stebbins, R. A. (1996) Volunteering: A Serious Leisure Perspective. *Nonprofit and Voluntary Sector Quarterly*, 25(2): 211–224.

Stebbins, R. A. (2005) Project-based Leisure: Theoretical Neglect of a Common Use of Free Time. *Leisure Studies*, 24(1): 1–11.

Uriely, N., Reichel, A. and Ron, A. (2003) Volunteering in Tourism: Additional Thinking. *Tourism Recreation Research*, 28(3): 57–62.

Wearing, S. (2001) *Volunteer Tourism: Experiences that Make a Difference*, New York: CABI.

Wearing, S. (2002) Re-centering the Self in Volunteer Tourism. In G. M. S. Dann (ed.) *The Tourist as a Metaphor of the Social World*, Wallingford: CABI, pp. 237–262.

Wearing, S. and Deane, B. (2003). Seeking Self: Leisure and Tourism on Common Ground. *World Leisure*, 1: 6–13.

Wearing, S., DeVille, A. and Lyons, K. (2008) The Volunteer's Journey Through Leisure into the Self. In K. D. Lyons and S. Wearing (eds) *Journeys of Discovery in Volunteer Tourism: International Case Study Perspectives*. Wallingford: CABI, pp. 63–71.

Wearing, S. and Wearing, B. (2001) Conceptualizing the Selves of Tourism. *Leisure Studies*, 20: 143–159.

Zahra, A. and McIntosh, A. J. (2007) Volunteer Tourism: Evidence of Cathartic Tourist Experiences. *Tourism Recreation Research*, 32(1): 115–119.

6 Developing and promoting sustainable volunteer tourism sites in Sabah, Malaysia

Experiences, dimensions and tourists' motives

Jennifer Kim Lian Chan

Introduction

The concept of volunteering while on vacation started in the early twentieth century (Wearing and Neil, 2001; Beigbeder, 1991) due to a growing awareness of anti-globalisation and environmental degradation issues in contemporary Western cultures (Uriely, Reichel and Ron, 2003). In the same vein, people from most parts of the world have become more educated in relation to environmental degradation and the negative consequences of economic progress. In fact, most Western governments are generally moving towards more sustainability initiatives, such as promoting green environments and the purchasing of environmental responsible goods and services, to reduce harmful effects on human health and the environment (Government of Canada, 2006). This is further supported by Drucker (1995: 255–257) who pointed out that 'a third/non-profit/ social sector', or volunteer organisations, will play a major role in the twenty-first century in order to sustain the smooth running of a civil society. At the same time, the Ecovolunteer Programme (2000) identifies the specific prominent features used to describe volunteer tourism, which includes: meeting the special interests of the volunteer tourist (wildlife, education, authentic experience, participatory in nature, opportunities to meet and work with like-minded people) and his/her purposes or reasons to travel; the participants' support towards the activities of the host (nature conservation or community development); and making new friends, learning new things and developing new skills. Volunteer tourism programmes contribute towards individual personal development and have a positive direct impact on social, natural and/or economic environments (Wearing, 2001) and sustainability. Thus, it is clear that a sustainable volunteer programme enables tourists to engage in voluntary activities while on vacations and the experiences contribute to their personal development and have an impact positively and directly on the social, nature and economic environments in which they participate (Responsible Travel Handbook, 2006). This implies that the experience dimensions of volunteer tourism are derived from different sources. An insight into the experience dimensions of volunteer tourism at a particular site will enhance the tourist satisfaction and marketing of the particular volunteer site (Crabtree and Gibson, 1992; Khan and Johnstone, 1995). Likewise, the successful implementation of sustainable tourism requires cooperation by a wide range of different stakeholders such as the public sector, different subsectors of businesses in the tourism industry, tourist attractions/sites, tourists and host communities (Vernon *et al.*, 2005; Aesh and Chan, 2008).

Literature documents a growing demand for participatory voluntary projects in areas such as educational learning, environmental or wildlife conservation works (Ellis, 2003) and community-based activities (Emmons, 2006). These participatory voluntary projects have been implemented in a few regions of Sabah, Malaysia since the1990s (to be discussed later). There has been an increase in empirical work related to volunteer tourism which has given rise to the concept and label volunteer tourism (Uriely *et al.*, 2003) particularly in areas of volunteer tourist experiences (Broad, 2003; Halpenny and Caissie, 2003). It is argued that tourist experiences are important issues to be considered when developing and promoting a sustainable volunteer tourism programme (Aabo, 2006). Nevertheless, there seems to be limited empirical work that specifically focuses on the experience dimensions from the tourist and host perspectives in sustaining volunteer tourism sites or projects. It is likely that volunteer tourists may not be able to search or identify suitable types of experiences that match individuals' interests and preferences, hence the need to investigate the motivational attributes of the tourists and their experiences including host communities in developing and promoting sustainable volunteer tourism programmes, which has been raised and further supported by Broad (2003). The understanding experience dimensions of volunteer tourists could contribute towards a more practical approach to develop and promote volunteer tourism sites via sustainable volunteer tourism programmes as well as conservation of natural resources by the tourists and project sites.

There has been a growing recognition of the importance of tourists' experience in various tourism sectors and empirical research on tourists' experiences has been conducted in a variety of areas, such as museums (Rowley, 1999), river rafting (Arnould and Price, 1993; Fluker and Turner, 2000), skydiving (Lipscombe, 1999), heritage parks (Prentice *et al.*, 1998) and heritage sites (Masberg and Silverman, 1996; McIntosh, 1999) and ecotourism experiences (Chan and Baum, 2007a; 2007b). It is postulated that volunteer tourists' experiences are critical in building sustainable volunteer tourism, as it is argued that volunteer tourism includes both service experience and service quality just like the tourism industry. Experience is regarded as an important element in influencing satisfaction and also the sustainability of volunteer tourism. The volunteer tourism experience is derived from the personal/emotive (tourists) and the utilitarian/functionality of the service performance (destination attributes). The psychological environment – the subjective personal reactions and feelings experienced when consuming a service – has been found to be an important part of consumer evaluation of and satisfaction with services (Otto and Ritchie, 1996). In particular, understanding experiential phenomena is crucial, as emotional reactions and decisions often prevail amongst consumers (Wakefield and Blodgett, 1994). It is argued that tourist experiences are individualistic, subjective and emotional in nature, while the service experience is inherently interpretive, subjective and affective (McCallum and Harrison 1985; Parasuraman *et al.*, 1988). The affective component of the service experience has been shown to consist of subjective, emotional and highly personal responses to various aspects of the service delivery. Otto and Ritchie (1996) develop six construct domains for the service experience, as shown in Table 6.1. It is argued that the construct of domains of service experiences can be extended to understand the volunteer tourist experience.

Elsewhere, Aabo (2006) suggests that sustainability in volunteer tourism programmes and sites infers the generation of economic benefit for the community, the conservation of the ecological environment and the respect of the host community culture. The provision of relevant volunteer experiences via the programme or sites is deemed vital in providing relevant volunteer tourist experiences and developing and promoting volunteer tourism sites in a more sustainable manner. This means that volunteer tourist experiences are related to the sustainable volunteer tourism that should incorporate the facilitation of host community ownership and control (Wearing, 2001). Thus, a sustainable volunteer tourism site enables tourists to engage in voluntary activities while on vacation, and this experience would contribute both to their own personal development and also have a positive and direct impact on the social, nature and economic environments in which they participate. Subsequently, positive relationships exist among the volunteer tourists, tourism operators and natural resource managers to establish better interpretations by guides (Talbot and Gould, 1996). Brown and Lehto (2005) propose two types of volunteer tourists known as: a) volunteer-minded who seek out opportunities that support their altruistic tendencies; b) vacation-minded who choose their volunteering location based on vacation advertising and promotion materials. It is clear that these two mindsets have different motives, values and decision-making approaches.

Table 6.1 The construct domains of service experience

Dimension	Examples
Hedonic	Excitement, Enjoyment, Memorability
Interactive	Meeting people, Being part of the process, Having choice
Novelty	Escape, Doing something new
Comfort	Physical comfort, Relaxation
Safety	Personal safety, Security of belongings
Stimulation	Educational and informative, Challenging

Source: Otto and Ritchie, 1996: 169

Volunteer tourism in Sabah and study sites

Sabah is a state of Malaysia, situated in the north-east of the island of Borneo, bordered by the South China Sea, Sulu Sea and the Celebes Sea. Sabah is a premier nature, adventure and cultural destination and is well known for its eco-tourism due to its rich biodiversity both on the land and in the sea. Since 1995, Sabah has become an attractive tourist destination for investors in the accommodation sector. Indeed, Sabah is a premier tourist destination for nature, cultural and adventure tourism due to its rich natural resources and protected areas. There are thirteen wildlife-based eco-tourism opportunities under the Sabah Wildlife Department. Hence, Sabah appeals as an 'activity-based' destination to modern travellers who are seeking activity-based attractions as opposed to 'destination travel' (Sabah Wildlife Department, 2008: 2–3).

The Sabah state government has been very active in running conservation programmes and projects together with private and non-profit organisations such as WWF (World Wide Fund for Nature), United Nations Development Project and Danish Co-operation to promote sustainable multiple use of natural resources, to conserve bio-diversity and the ecosystem and to protect indigenous habitats and wildlife (Sabah Wildlife Department, 2008). These efforts alleviate any abuse of rich tourism resources and encourage sustainable tourism development in the long run (Payne, 1997). The valuable sustainability and diversification perspectives are paramount to the sustainable tourism growth in the context of a growing tourism industry and conservation and preservation of tourism natural resources. In addition, the distribution of benefits is an important area and the host communities must grasp distinct benefits from volunteer tourism (Wearing, 2001; McGehee and Santos, 2005).

Review of the four volunteer tourism sites in Sabah, Malaysia

Over the last few years, there has been a steady increase in volunteer tourist arrivals to Sabah as a result of the visibility of website promotions and demand from tourists. Over the years, there has been a steady increase in the number of volunteer tourists, for example four volunteer tourism sites in Sabah, namely the English Teaching Project at Kota Marudu; Low Kawi Wildlife Park at Penampang; Pulau Tiga/Survivor Island and the Orangutan Rehabilitation Centre at Sandakan.

English Teaching Project at Kota Marudu, Sabah

The English Teaching Project as one of the volunteer tourism projects is made available in rural village environment in the northern part of the Sabah state, known as Kota Marudu. The two schools involved with volunteer travellers are the Seventh Day Adventist (SDA) and a Christian group. The range is from kindergarten, through to primary and secondary. The duration for volunteer involvement is about 10–15 lessons per week. The volunteer primarily acts as an assistant to the English teacher helping to take small groups and practices in conversational English. The volunteers also take charge of the classes and are involved in other areas such as sport, drama and music depending on the interest of the volunteers as well (Travellers Worldwide, 2008b: 1). In addition, volunteer tourists may chose to assist in science and maths as these subjects are taught in English. Volunteers do not need to have special qualifications since the idea of volunteering means to provide any help possible. This could mean enabling children to hear English pronunciation by native English speakers; providing additional help with other academic subjects or assisting in how to use a computer. The emphasis is on much needed language development and additional learning associated with cultural contact. The children are exposed to valuable experiences in learning, seeing and interacting with Westerners. Schools are keen to establish good working relationships with volunteers as it is felt that volunteers are one of the best ways to increase the standard of their students' English. It is important for the volunteers to enjoy themselves and to visit Sabah's attractions in addition to carrying out their volunteer jobs. Volunteer travellers are given familiarisation about the country and local cultures prior to the placement tourist sites (Travellers Worldwide, 2008b: 1–4).

Animal Care Project at Low Kawi Wildlife Park

The Animal Care Project in Low Kawi Wildlife Park is the second volunteer tourism project in Sabah, Malaysia. The park is located in a 280-acre forest reserve in the district of Penampnag and run by Sabah Wildlife Department. It is the home to many of Sabah's native and threatened species and has been a 'practice site' for education and fostering the awareness of conservation issues. The park was built with the aims of: increasing public awareness, providing world class animal facilities and enclosures, developing an environmental education programme, developing a programme for the captive breeding of endangered species and promoting research activities. It is a mix between a traditional wildlife park and a zoo. Travellers Worldwide have worked closely with Sabah Wildlife Department and take pride in their successful partnerships. Several volunteers have joined the Travellers Worldwide orangutan project run by Sabah Wildlife Department in order to help with various projects set up by the wildlife park in the past. Travellers Worldwide have been involved in the design of the sun bear enclosure and the graphic design of the wildlife park's information pamphlet since 2003. In addition, there is a Travellers Worldwide donation that reflects their commitment in education and raising awareness about conservation. At the same time, a specific volunteer programme and activities are planned for volunteer visitors. The duration of a volunteer project is four weeks, all volunteers must attend an induction with the Travellers Worldwide organiser and all contact with animals is supervised by staff at the Wildlife Park. The requirement for participation in this project include: a high level of fitness; working as part of a team; taking part in physical work in hot and humid conditions; the ability to cope with a tropical environment; a genuine interest in wildlife and recognise the educational importance of having animals in a wild-life park and not just a rehabilitation programme.

More importantly, the volunteers have the opportunity to learn about animal care across a variety of species and hands-on contact with some of the animals that have been hand reared. The rotation system is implemented to allow volunteers to have opportunities to learn and care for the range of animals at the park. Volunteers will spend ten days caring for the elephants, five days with primates and reptiles and five days with birds and hoof stock. The main activities involved are animal care including the cleaning of enclosures thoroughly to ensure a clean environment, preparing food/milk and feeding, bottle feeding and washing. In addition, the volunteers maintain records and report to the veterinarian on the behaviour and eating pattern of the animals and signs of illness. Other activities include assisting with visitor interaction and safety, providing general information and talking to visitors (Travellers Worldwide, 2008c).

Diving Project at Pulau Tiga (Survivor Island)

Pulau Tiga consists of three islands; this protected area is located in northern Borneo and the South China Sea. It offers excellent wildlife and a healthy coral reef with crystal clear water. A diving placement at the resort's diving school gives diving enthusiasts the chance to gain experience in Malaysia's tropical waters. The volunteer tourists assist dive masters in running the resort's dive centre, develop diving skills by taking courses and have the opportunity to do a total of 60 dives which is a prerequisite of the dive master course. Volunteers are involved with both dry and wet activities. The dry activities include looking after the dive centre; general cleaning; handling customer enquiries; manning the dive centre; guiding and assisting new learners; filling up oxygen tanks; cleaning dive equipment; assisting in the movements of boats; conducting English classes and preparing the kayaks or jet skis to be ready for use by guests staying at the resort. The wet activities include: helping the instructor while conducting diving lessons; supporting and guiding learners; preparing diving gear for lessons; interacting with new learners and cleaning and maintaining the diving equipment. Other experiences for volunteer tourists are monitoring and surveying the artificial reef which includes replacing and or repairing damaged coral; continuing the development of the reef; removing underwater rubbish; collecting and replanting coral fragments; monitoring and surveying the house reef; conducting an inventory of the marine life at the designated area; feeding fish, reporting on the quality of water; and developing an artificial reef, from the foundation of planting corals to setting up coral nursery units. These activities involve other volunteers who are in the placement and there is no supervision by a member of staff from the dive centre. The volunteer tourists need to sign up to open and advance water courses for the first month; emergency first aid response course and rescue divers course in the second month and a dive master course (optional) in the third month. Volunteer tourists need to bring their own dive equipment. Volunteers have the opportunity to engage in other activities, i.e. trails, active mud volcanoes and have the use of non-motorised water sports equipment, i.e. kayaks (Travellers Worldwide, 2008d).

Orangutan Rehabilitation Project

The Orangutan Rehabilitation Project is located at Sepilok Orangutan Rehabilitation Centre, Sandakan, and Sabah, Malaysia. The centre is situated on the east coast of Sabah and is run by Sabah Wildlife Department; it is the most popular site amongst the volunteer tourism sites in Sabah. The centre has enormous areas of virgin rainforest and a wealth of indigenous plants, animals and birds. The wildlife reserve contains the orangutan rehabilitation centre. The centre rescues abandoned, injured or orphaned orangutans and liberates them from captivity. Animals are treated for injuries and are nursed back to health. The rehabilitation program aims to return captive animals to the wild. The centre opened in 1964 and has successfully rehabilitated over 100 orangutans. The centre aims to provide public education on conservation, research and assistance to other endangered species. The orangutan rehabilitation is a two-month project and is very popular amongst volunteer tourists worldwide. The volunteer tourists learn about orangutan husbandry management by working in both the indoor and outdoor nursery and orangutan paediatric and veterinary care. Volunteers also conduct a field survey on the orangutan population in Speilok Reserve, a nocturnal animal survey and an extended orangutan and a nocturnal animal survey in areas deeper into forest reserve. The indoor nursery looks after the infants from a few months old to around four years old. The volunteer tourists have the opportunity to transfer infant orangutans from the sleeping cages to the playing cages; clean the cages and indoor nursery areas; prepare food; feed the babies; distribute bananas and see to the sick infants. The outdoor nursery is the home to juveniles, from five to eight years old.

The volunteer tourists' duties involve: sweeping leaves from surrounding paths and the cleaning of nursery areas; transporting the bananas from the storeroom to the nursery; feeding bananas to orangutans at the outdoor nursery; assisting the rangers to carry bananas to the tourist platform and completing observation sheets on the orangutans which use the feeding platform. This means volunteer tourists have plenty of time for observation of the orangutans in their natural surroundings. Other duties include: assisting the veterinarian in the treatment of sick orangutans, assessing the daily health status of any sick orangutan; the recording of any treatment given; observing and collecting data on nocturnal animals within the reserve through direct sighting in day and night hours. It also involves enrichment, play and climbing exercise with the orangutans, which aims to contribute to the satisfaction of the volunteer as they have contributed something that is both hands on and conservation based. The work is done on a rotational basis with different groups of volunteers working within each of the different departments. The project allows involvement in the early stages of rehabilitating orphaned orangutans, learning how to handle them, teaching them to climb ropes and small trees and encouraging them to go into bushes to build nests. Volunteers for the placement must be animal/nature lovers or like being involved in conservation work. Volunteers are required to be adaptable to a different way of life and to work ethically and flexibly in the working routines outlined. The project is very physically demanding and volunteers should be prepared to get dirty (Travellers Worldwide, 2008e).

In summary, the review of secondary data of the four volunteer tourism sites in Sabah show that volunteer tourists are exposed to different volunteer activities and experiences. Thus, this implies that there may be different travelling motives for each of the volunteer tourism sites. At the same time it is possible that these volunteer tourists are also attracted to the key attractions in Sabah such as Mount Kinabalu and other ecotourism sites such as Lower Kinabatangan River, Maliau Basin.

Research method and data collection

This chapter describes an exploratory qualitative research project, designed to investigate key experience dimensions of volunteer tourists and volunteer site attributes of the four volunteer sites as well as the programme operator perspective in promoting and sustaining the volunteer tourism in Sabah. An inductive approach was adopted to address the research objectives, since they are subjective in nature within an interpretative paradigm. The data collection involved the review of secondary data with regard to the four volunteer tourism sites. The primary data was collected via in-depth interviews, with regards to the volunteer tourist experience gained, benefit sought and motives to participate in volunteer tourism in Sabah. Data is cross sectional and conducted on sites. In-depth interviews were conducted with five volunteer tourists at four volunteer tourism sites: the English Teaching Project at Kota Marudu; Low Kawi Wildlife Park at Penampang; Pulau Tiga/Survior Island and the Orangutan Rehabilitation Centre at Sandakan. A total of 20 respondents were interviewed.

The sample size was determined when information and theoretical insights reached saturation, which constitutes hearing the same information reported without anything new being added. In-depth interviews were conducted using tape recordings of 30–40 minutes' duration. In-depth interviews stopped at respondent No. 20. A total of five English-speaking respondents were interviewed at each volunteer tourism site. The sample respondents were international volunteer tourists who had participated in volunteer programmes at the four volunteer tourism sites in Sabah and only English-speaking volunteers were used. Simply, the international English tourists form the largest market segment for volunteer tourists in Sabah. The data was collected from February to April 2009. A convenience-sampling technique, based on the lists of volunteer tourists' arrivals and departure dates from the projects, was adopted as the sampling frame. This technique is a key feature of qualitative research and presents the typology of sampling strategies in a qualitative inquiry (Patton, 2002; Kuzel, 1992).

Data analysis

A total of 20 audio tapes from the in-depth interviews were transcribed verbatim in order to secure the authenticity and richness of the data from the four volunteer tourism sites. The data was analysed using a qualitative-phenomenological approach as the analysis technique, which was data and conceptually driven. The unit of coding was mostly based on a single phrase or several significant statements that were meaningful in order to generate themes related to the research questions. Key themes and patterns emerged from the coding process. The interview responses were read and re-read, and analysed by drawing out the key themes and variables relating to the descriptive themes – volunteer tourist motives and key experiences gained, volunteer tourism site attributes from the perspectives of the staff at project centres and economics gained for the operators, as presented in Table 6.2. Thus, the themes are categorised based on the components – volunteer tourists and tourism sites as reviewed in the literature. The responses of volunteer tourists' experiences were further analysed by using the construct domains of service experiences by Otto and Ritchie (1996) as presented in Table 6.1. Responses were further categorised as personal and functionality dimensions.

Findings and discussion

The empirical evidence points out that the type of volunteer tourists to Sabah can be categorised as 'vacation-minded' types as the majority of them tend to choose their volunteering location based on vacation advertising and promotion materials, while a small number of them tend to choose their vacation based on seeking opportunities that support their altruistic tendencies, known as 'volunteer-minded'. This is evidenced from the interview responses that some of the volunteers indicated that their participation of volunteer programmes makes a positive contribution to the project and their personal development rather than just a vacation/leisure activity (seen later in Table 6.3).

The volunteer tourists' experiences though participating in volunteer tourism in Sabah

The empirical evidence shows a range of key experience dimensions were derived from the volunteer tourist verbal responses based on the four volunteer tourist sites. The underlying key experience dimensions gained by the volunteer tourists in participation of volunteer tourism activities in Sabah, Malaysia gained from the interview responses show that a wide range of experiences are gained in participating in volunteer tourism, presented in Table 6.2. The evidence seems to suggest that volunteer experiences are derived from the emotive aspects of the tourists which are related to hedonic, stimulation, novelty; and interactive and functionality aspects of the volunteer tourism which are related to the wildlife and conservation projects – reef and orangutan.

Table 6.2 Volunteer experience themes that emerged from interview responses

Service experience constructs (Otto and Ritchie, 1996) (affective, emotive aspects)	Key experience dimensions (personal and functionality)	Some of the volunteer tourist verbal responses at the four volunteer tourism sites
Hedonic: excitement, enjoyment	Personal dimension: education and learning and life-changing experience	Is an exciting placement that offers great hospitality (Respondent 1, male, Kota Marudu)
		I enjoy teaching English to English teachers and in return I learn Malay and Malaysia's lifestyles (Respondent 2, female, Kota Marudu)
Stimulation: education, informative and challenging	Functionality dimension: wildlife	I have opportunity to contact with some rare species of wildlife that I have dreamed before (Respondent 6, female, Low Kawi Park)
	Personal: learning	I have the opportunity to learn about animal care across a variety of species; personally interact with some of the animals that have been hand reared (Respondent 7, male, Low Kawi Park)
	Interactive	
	Learning experiences	I have learnt how to take care of elephants, primates and reptiles and birds ... it is a challenging experience for me (Respondent 8, male, Low Kawi Parkt)
Novelty: escape, do something new	Personal dimension: novelty	it is something very different for me the coral reef and its surrounding ...as diver I have the chance to expose to both dry and wet activities ... (Respondent 11, male, Pulau Tiga)
	Functionality dimension: reef conservation	it nice to be able to assist the diver masters and help in reef conservation work ... (Respondent 12, female, Pualu Tiga)
	Skills and conservation activities	I felt it was a perfect place to practice volunteer work and have holiday at the same time ... this place is excellent for the divers to pick up diving skills and diving activities and participate in coral reef conservation (Respondent 14, female, Palau Tiga)

continued

Interactive: meeting people, being part of the process, having choice	Personal dimension: interactive learning – rehabilitation of orangutan	I have the opportunity to be part of the team to rehabilitate orphaned orangutans and learn how to handle them, teach them to climb ropes, small trees and encourage them to go into bushes to build nests (Respondent 16, male, Sepilok Orangutan Rehabilitation Centre)
	Being part of the process	To work this close to orangutans is an experience in itself (Respondent 17, male, Sepilok Orangutan Rehabilitation Centre)
	Meeting people of different cultures	The project has given me the chance to work with many different people some of whom were from a different culture (Respondent 18, female, Sepilok Orangutan Rehabilitation Centre)
	Culture experiences	Learned to live and work with other volunteers and made some very good friends, learned basic Malay language, learned about their country and culture, visited many places and tried out local cuisine (Respondent 18, female, Sepilok Orangutan Rehabilitation Centre)
	Wildlife conservation, interactive	Worked alongside members of staff who have a lot of knowledge in wildlife conservation and also helped out in feeding the orangutan babies. (Respondent 9, female, Sepilok Orangutan Rehabilitation Centre)
	Learning conservation	The best part was learning that the main threat to orangutans is from human activity, in the form of destruction of their habitat for logging and conversion of forests to palm oil plantations and farmlands. There is a need for conserving the orangutans and its rainforests. (Respondent 20, female, Sepilok Orangutan Rehabilitation Centre)
		To me, the best part is learning about organutans and having the opportunity to enjoy and explore Sabah; especially the moment we reached the summit of Mount Kinabalu in time to catch the sun rise (Respondent 17, male, Sepilok Orang Utan Rehabilitation Centre)

The following section presents some of the evidence of the themes that emerged from in-depth interviews.

In the case of the English Teaching Project at Kota Marudu, the responses from the interviews showed hedonic (excitement and enjoyment) being the personal experience dimension, as reflected in the responses: 'Is an exciting placement that offers great hospitality' as pointed out by respondent 1, and 'I enjoy teaching English to English teachers and return I learn Malay and Malaysia's lifestyles', as stated by respondent 2.

Interestingly, in the case of the Animal Care Project at Low Kawi Wildlife Park, 'learning interactive' as a personal experience dimension and 'wildlife' as a functionality dimension emerged as two key experience dimensions from the responses. These are evidenced from the interview responses: 'I have opportunity to contact with some rare species of wildlife that I have dreamed before' as pointed out by respondent 6. The wildlife aspect is highlighted by respondent 7: 'I have the opportunity to learn about animal care across a variety of species; personally interact with some of the animals that have been hand reared'.

In the case of the Diving Project at Pulau Tiga: 'stimulation and reef conservation' are the key experience dimensions emerging from the responses. The 'novelty' is evidenced from respondent 11 who stated that 'it is something very different for me – the coral reefs and its surroundings … as a diver I have the chance to be exposed to both dry and wet activities'; whereas the reef conservation aspect is pointed out by respondent 12 whereby 'it is nice to be able to assist the divermasters and help in reef conservation work'.

For Sepilok Orangutan Rehabilitation Centre, it was found that interactive and wildlife conservation emerged as key experience dimensions, as reflected in the following responses:

> I have the opportunity to be part of the team to rehabilitate orphaned orangutans and learn how to handle them, teach them to climb ropes, small trees and encourage them to go into bushes to build nests.
>
> (Respondent 16)

> To work this close to orangutans is an experience in itself. To me, the best part includes learning about orangutans and having the opportunity to enjoy and explore Sabah; especially the moment we reached the summit of Mount Kinabalu in time to catch the sunrise.
>
> (Respondent 17)

> The project has given me the chance to work with many different people some of whom were from a different culture … to learn to live and work with other volunteers … I made some very good friends, learned basic Malay language, learned about their country and culture, visited many places and tried out local cuisine.
>
> (Respondent 18)

> Worked alongside members of staff who have a lot of knowledge in wildlife conservation and also helped out in feeding the orangutan babies.
>
> (Respondent 19)

The best part was learning that the main threat to orangutans is from human activity, in the form of destruction of their habitat for logging and the conversion of forests to palm oil plantations and farmlands. There is a need for conserving the orangutans and its rainforest.

(Respondent 20)

Evidently, these responses seem to suggest that the key experience gained from participating in volunteer tourism is derived from the personal and service performance of the volunteer tourism. This means that the emotive aspects of the volunteer tourist and the functionality of volunteer sites are vital in promoting and developing sustainable volunteer tourism. Put simply, the experience gained by the individual volunteer tourists can become important word-of-mouth marketing for the volunteer tourism site; and similarly, the volunteer tourism site attributes (what attracts the volunteer tourists – activities/project, wildlife and conservation projects) are key motives for the volunteer tourists to participate in the volunteer tourism. This seems to suggest that the volunteer tourists' motives and volunteer tourism site (activities and project) are regarded as two important aspects that contribute to volunteer tourism experiences; and subsequently these influence the promotion and sustainable development of volunteer tourism.

The volunteer tourists' motives through participating in volunteer tourism in Sabah

Likewise, the empirical evidence also suggests a range of volunteer motives in participating in volunteer tourism in the four volunteer tourism sites, as presented in Table 6.3. This seems to suggest that the motives of volunteer tourists are related to meeting the special interests of the volunteer tourist: wildlife, education, authentic experience, participatory in nature, opportunities to meet and work with other like-minded people and his/her purposes or reasons to travel; the participants' support towards the activities of the host (nature conservation or community development); making new friends; learning new things and developing new skills. Thus, it seems to conclude that the participation of volunteer tourism programmes must contribute towards individual personal development and have a positive direct impact on social, natural and/or economic environments (Wearing, 2001) and sustainability. These echoed the findings of educational learning, environmental or wildlife conservation works (Ellis, 2003) and community-based activities (Emmons, 2006).

Table 6.3 Volunteer tourists' motives that emerged from the interview responses

Volunteers' motives	Volunteers' responses
Improve knowledge on orangutan conservation	To understand orangutans and practice and contribute to conservation by working very closely with them. (Respondent 17, male, Sepilok Orangutan Rehabilitation Centre)
Involve in physical activities or 'hands on' work	I would like to have hands on experience of working and taking care of animals – elephants, orangutan. (Respondent 7, male, Lok Kawi Park)
Fulfil the dream in teaching English	To share and to work with students and teachers in improving their English has been my dream for several years. (Respondent 4, female, English Teaching Project at Kota Marudu)
Personal or career development	This experience will help me with my course at the university and future career. (Respondent 9, female, Low Kawi Park)
Culture experience	I want to experience living in another culture. I heard Sabah is one of the perfect places. (Respondent 3, male, English Teaching Project at Kota Marudu)
Meet people with the same interest	I want to work with people with the same interest and also learn from them as well. (Respondent 18, male, Sepilok Orangutan Rehabilitation Centre)
Being a responsible traveller and have fun	I want to be a responsible traveller and to have a lot of fun in Sabah… (Respondent 14, female, Diving Project at Pulau Tiga)
Gain a life changing experience	Participating in volunteer tourism in Sabah will be life changing experience that I gained will be an achievement for me (Respondent 12, female, Diving Project at Pulau Tiga)
Make a positive contribution to the project	To gain personal experience on conservation of wildlife. (Respondent 20, female, Sepilok Orangutan Rehabilitation Centre)
	I hope to contribute to the conservation project by sharing my knowledge, skills, teamwork and energy. (Respondent 14, female, Diving Project at Pulau Tiga)

The evidence suggests that the volunteer tourist experiences are derived largely from the personal/emotive aspects and also the main activities participated in. Thus, it seems to suggest that for sustainable long-term development and effective promotion, it is vital to take into consideration volunteer tourism experiences gained and their dimensions, as these elements affect the satisfaction of the volunteer tourists and meeting their motives. Clearly, international volunteer tourism is a sustainable form of tourism that generates wealth and benefits in many different ways. Predominantly, it provides volunteers' time, skills, knowledge and financial support to community project in regions of the world. In fact, the volunteer tourists can provide great assistance to local schools or communities with their skills, new ideas and materials, to heritage sites and parks or environmental awareness campaigns, or to the local employment by increasing the demand for local goods and services or lending expert knowledge and skills to training programmes. This implies that volunteer tourists are more knowledgeable, socially and environmentally aware and responsible which means they inflict less damage on the environment than the average tourist (Emmons, 2006). To some extent, volunteer tourism enables the volunteer tourists to understand how global concerns are felt locally, which may ultimately lead to a permanent shift to a more sustainable lifestyle as a result of their personal desire and motives through the experienced gained. In this regard, volunteer tourists tend to be more educated, want a life-changing experience while on holiday and show they are keen to contribute their time and energy to the local community and be transformed by the experience, as evidence from the interview responses in Tables 6.2 and 6.3. Often, volunteer tourists are seeking something different and they are often very well aware of global issues through personal research and experiences from their own countries. They are often inspired to work abroad in different countries and are willing to give their time and energy in productive ways during their holidays. The finding of volunteer tourist experiences and motives in Sabah confers with the features used to describe volunteer tourism by the Ecovolunteer Programme (2000), which includes: meeting other volunteer tourists with similar interests (wildlife, orangutan conservation project, education, gaining authentic experience, participatory in nature), supporting the activities of the host (nature conservation or community development); making new friends, learning new things and life-changing experience as well as making positive contributions to conservation projects. Hence, volunteer tourists are seeking opportunities to learn new cultures and gaining a life-changing experience, meeting people with similar interests and having fun. More importantly, they are seeking opportunities to make positive contributions to conservation projects or improvements to local communities in many different ways. As for volunteer tour operators, attractive returns and retaining economic benefits locally, whilst attracting the right attitude of volunteer tourists (respecting the local culture, open-mind, sensitive to environmental issues and nature conservation) is important.

Conclusion and contribution

The exploratory research aimed to examine volunteer tourists' underlying experience dimensions of volunteer tourists for the four volunteer tourism sites in Sabah, Malaysia. Subsequently, it highlighted the key elements to be considered in promoting and developing sustainable volunteer tourism in Sabah by taking into account the personal and functionality aspects of volunteer tourism. The findings reveal that volunteer tourist experiences are attributed to the personal emotively aspects and wildlife conservation efforts seem to play a significant functionality of the tourist experiences. It shows that the volunteer tourist experience dimensions are multidimensional. The key experience dimensions that relate to personal emotively aspects are hedonic, stimulation, novelty and interactive, whilst wildlife, wildlife conservation, coral reef, group members/host/staff are regarded as key experience dimensions attributed to the personal experiences and also can be considered as functionality of the volunteer tourism sites. Clearly, the dimensions of volunteer tourist experiences are produced by the integration of a combination of factors – personal and volunteer tourism service performance (functionality) – and that each volunteer tourism site offers different element of service experience dimensions as emerged from the empirical evidence. Based on the volunteer tourists' motives, it seems to conclude that volunteer tourists to Sabah can be categorised as 'vacation-minded' who are motivated to participate in volunteer tourism due to thier desire to learn about orangutan conservation, involvement in physical activities or 'hands on' work and fulfil the dream to work with orangutans as well as personal or career development. This finding is consistent with the finding of Aesh and Chan (2008).

In order to satisfy the varying wants of potential volunteer tourists or customers, volunteer tourism operators need to emphasize volunteer tourism activities and personal learning and development, level of involvement of tourists and local culture experiences, in which they have engaged physically at the sites and their natural environment, their interaction with the site service staff and group members as well as the learning and information acquired during the visit. This suggests that the promotion and development of volunteer tourism sites may need to pay more attention to their surrounding natural resources, site project, service staff and the interactive between the tourists and staff/host as well as information about wildlife and local culture. These factors contribute significantly to the volunteer tourists' experience. The volunteer tourism site, its specific project, its attraction and human interaction of that particular site contribute to the tourist's experience and later can be an important selling and promotion point for that site. Equally important is the volunteer program that allows full participation in the activities and projects in the respective sites will generate significant experiences for the tourists. Thus, having a unique volunteer program that generates memorable experience is crucial for long-term development. The tourist experience that generates the hedonic, interactive, novelty, stimulation and comfort seems to be important for the volunteer tourism in addition to the meeting of the individual motives.

The chapter summarises the key experience dimensions of volunteer tourism and the key motives in participating volunteer tourism in Sabah, Malaysia. These key experience dimensions and motives may implicate the development and promotion of volunteer tourism sites. The chapter proposes that, for sustainable volunteer tourism site development and promotional framework, one needs to consider the key experience dimensions of volunteer tourists, meeting their motives and the key activities/project of volunteer tourism sites that will allow the participation of volunteer tourists at its fullest. Such a framework may serve as a useful guide for other sites in term of developing, managing and promoting sustainable volunteer tourism in a more competitive manner. The findings also suggest that further research in generalisation of the experience dimensions in volunteer tourism in other sites.

Reference

Aabo, D. (2006) *Sustainable Tourism Realities: A Case for Adventure Service Tourism.* Available from http://www.xolaconsulting.com/Aabo_David_Adventure_Service_Tourism.pdf. Accessed 28 June 2008.

Aseh, A. and Chan, J. K. L (2008) *Developing a Sustainable Volunteer Tourism Using Action Research: An Exploratory Study at Sepilok Orangutan Rehabilitation Centre, Sabah* in RARC International Tourism Conference 2008, co-organized by Rikkyo Amusement Research Centre (Rikkyo University, Japan); Erasmus University, Holland and Universiti Science Malaysia, 3–4 November, USM, Penang.

Arnould, E. J. and Price, L. L. (1993) River magic: extraordinary experience and the extended service experience. *Journal of Consumer Research*, 20 (1): 24–45.

Beigbeder, Y. (1991) *The Role and Status of International Humanitarian Volunteers and Organizations.* Netherlands: Martinus Nijhoff Publishers,

Broad S. (2003) Living the Thai life – a case study of volunteer tourism at the Gibbon Rehabilitation Project, Thailand. *Tourism Recreation Research*, 28(3): 63–72.

Brown, S. and Lehto, X. (2005) Travelling with a purpose: understanding the motives and benefits of volunteer vacationers. *Current Issues in Tourism*, 8 (6): 479–496.

Chan, J. K. L. and Baum, T. (2007a) Ecotourists' perception of ecotourism experience in Lower Kinabatangan, Sabah, Malaysia. *Journal of Sustainable Tourism*, 15(5): 574–590.

Chan, J. K. L. and Baum, T. (2007b) Motivation factors of ecotourists in ecolodge accommodation: the push and pull factors, *Journal of Asia Pacific Travel Tourism*, 12(04): 349–364.

Crabtree, A. and Gibson, A. (1992) A case-history symbiosis between reef tourism, education and research. In B. Weiler (ed.) *Ecotourism: Incorporating the Global Classroom*, 1991 international conference papers, University of Queensland, Brisbane, Australia. Bureau of Tourism Research, Canberra, pp. 217–221.

Drucker, P. F. (1995) *Managing in a Time of Great Change*, New York: Truman Tallye Books/Plume, pp. 255–257.

Ecovolunteer Programme (2000) http://www.ecovolunteer.org. Accessed 26 June 2008.

Ellis, C. (2003) Participatory environmental research in tourism, a global view. *Tourism Recreation Research*, 28(30): 45–55.

Emmons, K. (2006) Meaningful tourism: educational travel and voluntourism. *Tourism Authority of Thailand – E-Magazine.* Available from http://www.tatnews.org/emagazine/2874.asp. Accessed 31 July 2008.

Fluker, M. R. and Turner, L. W. (2000) Needs, motivations, and expectations of a commercial white water rafting experience. *Journal of Travel Research*, 38: 380–389.

Government of Canada (2006) Policy on green procurement http://www.tpsgc-pwgsc.gc.ca/ecologisation-greening/achats-procurement/document/politique-policy-eng.pdf.

Halpenny, E. A. and Caissie, L. T. (2003) Volunteering on nature conservation projects: volunteer experience, attitude and values. *Tourism Recreation Research*, 28(3): 25–33.

Khan, B. and Johnstone, G. (1995) *Ecotourism Based Reef Research and Education: The Acropora Health Monitoring and Reef Operator Logbook Programs.* Alice Springs: Ecotourism Association of Australia.

Kuzel, A. J. (1992) *Sampling in Qualitative Inquiry.* California: Sage.

Lipscombe, N. (1999) The relevance of the peak experience to continued skydiving participation: a qualitative approach to assessing motivation. *Leisure Studies*, 18(4): 267–288.

Masberg, B. A. and Silverman, L. H. (1996) Visitor experiences at heritage sites: a phenomenological approach. *Journal of Travel Research*, (4), 20–25.

McCallum, J. R. and Harrison, W. (1985) Interdependence in the service encounter. In J. A. Czepiel (ed.) *The Service Encounter: Managing Employee/Customer Interaction in Service Business.* Lexington, MA: Lexington, pp.35–48.

McGehee, N. and Santos, C. (2005) Social change, discourse, and volunteer tourism. *Annals of Tourism Research*, 32(3): 760–776.

McIntosh, A. (1999) Into the tourist's mind: understanding the value of the heritage experience. *Journal of Travel and Tourism Marketing*, 8(1): 41–64.

Otto, J. E. and Ritchie, J. R. B. (1996) The service experience in tourism. *Tourism Management* 17(3): 165–174.

Patton, M. Q. (2002) *Qualitative Research and Evaluation Methods* (3rd edn). California: Sage.

Parasuraman, A., Zeithaml, V. A. and Berry, L. L. (1988) SERVQUAL: a multiple-item scale for measuring consumer perception of service quality. *Journal of Retailing*, 64(1): 2–40.

Payne, J. (1997) *WWF Malaysia (Sabah).* Available from http://www.sabah.org.my/bi/know_sabah/environment_conservation.asp. Accessed 3 July 2008.

Prentice, R. C., Witt, S. F. and Hamer, C. (1998) Tourism as experience – the case of heritage parks. *Annals of Tourism Research*, 25(1): 1–24.

Responsible Travel Handbook (2006) Available from http://www.transitionsabroad.com/listings/travel/responsible/responsible_travel_handbook.pdf. Accessed 10 January 2008.

Rowley, J. (1999) Measuring total customer experience in museums. *International Journal of Contemporary Hospitality Management*, 11(6): 303–308.

Sabah Wildlife Department (2008). Available from http://www.sabah.gov.my/jhl/. Accessed 28 June 2008.

Talbot, B. and Gould, K. (1996) *Emerging Participatory Monitoring and Evaluation Programs in Two Ecotourism Projects in Peten, Guatemala.* USA: Yale School of Forestry and Environmental Studies, Yale University.

Travellers Worldwide (2008a) *Voluntary Projects.* Available from http://www.travellersworldwide.com. Accessed 28 June 2008.

Travellers Worldwide (2008b) *Travellers: Teaching Projects on Borneo Island Malaysia – Additional Information*, p. 1–4; Issues 1 and 2 (online). Available from http://www.travellersworldwide.com. Accessed 29 June 2008.

Travellers Worldwide (2008c) *Conservation: Lok Kawi Wildlife Park Malaysia. Malaysia – wildlife-park additional-info-travellers.doc-Issue 2 (Online).* Available from http://www.travellersworldwide.com. Accessed 29 June 2008.

Travellers Worldwide (2008d) *Work Experience: 'Survivor' Island Diving Project Malaysia –Malaysia-Diving-Pulau-Additional –Info-Travellers.doc-Issue 1 (Online).* Available from http://www.travellersworldwide.com. Accessed 29 June 2008.

Travellers Worldwide (2008e) *Conservation: Sepilok Orangutan Rehabilitation Malaysia-Malaysia-Orangutan –additional-info-travellers.doc.-Issue 1 (Online).* Available from http://www.travellersworldwide.com. Accessed 29 June 2008.

Uriely N., Reichel A. and Ron, A. (2003) Volunteering in tourism: additional thinking. *Tourism Recreation Research*, 28(3): 57–62.

Vernon, J., Essex, S., Pinder, D. and Curry, K. (2005) Collaborative policymaking: local sustainable projects. *Annals of Tourism Research*, 32(2): 325.

Wakefield, K. L. and Blodgett, J. G. (1994) The importance of servicescapes in leisure settings. *Journal of Services Marketing*, 8(3): 66–76.

Wearing, S. (2001) *Volunteer Tourism: Experience that Make a Difference.* New York: CABI Publishing,

Wearing, S. and Neil, J. (2001) Expanding sustainable tourism's conceptualization: ecotourism, volunteerism, and serious leisure, in Stephen F., McCool, R. and Neil Moisey (eds) *Tourism, Recreation and Sustainability.* Wallingford: CAB International, pp. 233–254.

7 Volunteer tourism as a life-changing experience

Anne Zahra

Introduction

This chapter looks at the motivation, experiences and the long-lasting impact on the lives of ten volunteer tourists who undertook their volunteering holidays between 1989 and 2000. This chapter highlights how volunteer tourism is 'as much a journey of the self as it is a journey to help others' (Wearing *et al.*, 2008: 63). The researcher worked alongside some of the volunteers on the volunteer projects they participated in and for some, the researcher was aware that volunteer tourism experience had a significant impact on them and their future choices and direction in life. The research approach can be classified as ethnographic case studies because of the small sample, detail and the in-depth and personal nature of the narratives. This chapter is a longitudinal study providing evidence of the volunteering experience leading to a change in life course. The changes are presented in the context of the motivations to undertake the volunteering experience, the experimental experience (Cohen, 1979) and the long-lasting changes in the years following.

Literature on volunteering is very broad. This study is situated within the research related to international volunteering. Weinmann (1983) and Carlson (1991) looked at personal development in relation to experiencing a new culture, and how it acts as a catalyst for volunteers to gain insights into alternative values, beliefs and ways of life. However, these studies did not explore the actual changes in the volunteers' lives and their long-term impact. Volunteer tourism has intrinsic rewards; the potential to change participants' perceptions about society, self-identity, values and their everyday lives. Indeed, previous volunteer tourism studies have reported how, from their volunteer tourism encounter, volunteers experience self-reflection, increased social awareness and support, and experience a subsequent change to their daily lives and belief systems (Arai, 2000; Broad, 2003; McGehee and Santos, 2005; Simpson, 2004; Stoddart and Rogerson, 2004; Wearing, 2001, 2002).

The primary motivations of the volunteers were to experience a culture, go overseas, go on a holiday in which everything was organised and do something worthwhile. When they were confronted with suffering, poverty, cultures embedded with deep values devoid of materialism and consumerism, combined with the cheerfulness of the host communities amid the lack of basic needs, each volunteer underwent a cathartic (Ryan, 1997; Zahra and McIntosh, 2007) and life-changing (Wearing, 2002) experience.

Methodology

The volunteer tourist's life-changing experience is examined through exploratory qualitative research. The study narrates ten young volunteer tourists' life-changing experiences. Their countries of origin were Australia and New Zealand. They were aged between 19 and 23 years, at the time they participated in the volunteer tourism project. They assisted in grass roots community projects, organised by secular NGOs, in developing countries such as the Philippines, India, Tonga and Fiji between 1989 and 2000.

The co-ordinating NGO of the volunteer programmes was an Australian non-government organisation, RELDEV, registered with the Australian Agency for International Development (AusAID) that provides community development projects in Asia and South America. Besides community development projects, it organises development education projects in which young volunteers from Australia and New Zealand, aged between 16 and 26 years, participate in three to four week welfare/development projects in developing nations in the Pacific and Asia. These welfare projects provide 'on the ground' assistance to communities and engage the volunteers and community in a mutual exchange. This organisation is typical of

> [a] range of NGOs [that] undertake programmes that focus on personal development potentiality in tourism, which, in the past, has not been characteristic of tourism organizations. These organizations and their projects seek to be locally identified and sustainable, while providing the tourist with an opportunity to learn and become involved in development issues. These projects incorporate many of the key elements that are considered to be essential to the underlying concept of alternative tourism.
>
> (Wearing, 2001: 209)

The volunteers participated in what Stebbins (2004) called project-based volunteering. Callanan and Thomas (2005) label the participants as shallow volunteer tourists because of the short duration and the nature of their work.

The author has been involved in volunteer tourism both as a participant and as a coordinator of volunteer tourism projects since 1988 and knew all the participants personally. The personal relationship and trust facilitated the participant in sharing their experience. In-depth unstructured interviews were undertaken with ten volunteer tourists aged between 25 to 42 years. The author travelled with six of the volunteer tourists interviewed. She knew they had a life-changing experience while participating in a volunteer tourism project in their youth. These interviews were conducted between 7 and 18 years after the volunteer tourism experience. The volunteer tourists came from affluent, urbanised, developed and secular societies. The volunteer tourist's anonymity is maintained and pseudonyms have been used in this chapter.

The methodological approach of this research investigation utilises testimonial narratives seeking to explore the deep, personal and experiential of the volunteer tourists. This approach seeks the 'voice that speaks to the reader through the text in the form of an "I" that demands to be recognized' (Beverly, 2000: 556). Narrative is distinct from discourse, it is retrospective, articulates meaning and expresses emotions, thoughts and interpretations (Chase, 2005). The researcher gives *voice* to the narrator to articulate their self, reality and experience rather than facts. Unstructured in-depth interviews were used to collect data. This study will refer to those interviewed as narrators not interviewees. 'The interviewee as a narrator is not an interest in the other's "authentic" self or unmediated voice but rather an interest in the other as narrator of his or her particular biographical experiences as he or she understands them' (Chase, 2005: 661). The author recognises that narratives are socially situated and interactive. The relationship between the researcher and the narrator influences and shapes the story being told. The narrator's voice dominates, controlling the flow of the interview.

Testimonial narratives focus on making the narrator's story heard with less importance given to the researcher's interpretative processes, therefore a distance is created between the researcher's and narrator's voices (Atkinson, 2002). For the purposes of this chapter the narratives were categorically analysed, a sentence or paragraph is extracted from narrative to highlight a theme deduced by the author (Lieblich *et al.*, 1998).

These narratives are not being presented with the aim of drawing generalised conclusions for all volunteer tourists. The range of narrative possibilities for the volunteer tourist experience are potentially limitless (Gubruim and Holstein, 2002) due to social, cultural and historical conditions influencing the person concerned. The value of these narratives is that they provide us with deep insights into volunteer tourism or other tourist experiences. 'Narrative theorists point out that narrative research is embedded in and shaped by broad social and historical currents, particularly the ubiquity of personal narratives in contemporary Western culture' (Chase, 2005: 669).

Life-changing narratives

Motivations for volunteer tourism

The common and primary motivation of all of the participants was to go overseas and visit another country rather than seeking a volunteer work experience. Travel to other countries for Australians and New Zealanders is termed overseas as they need to cross a sea to reach their destination. For some volunteers, they had never been overseas before and this was the main attraction. This can be demonstrated by the following:

Wiri: I had never been overseas before. It was an opportunity to see a country and
get to know the people.

Toni: I was already thinking about going with a friend on a holiday overseas. I had never been overseas before.

A secondary motivation common to all the respondents was 'seeking to do something worthwhile' and 'to help disadvantaged groups'. This motivation can be characteristic of idealism found in the years of one's youth. The following quote illustrates this point:

Aata: Around the time, a friend invited me to the Philippines service project. I think I was not too interested at first, I suppose I did not know much about it, I was hesitant. But what got to me or what really convinced me was this same friend who said to me 'You can go on a holiday, spend money and do nothing or you can go to the Philippines and help people'. This statement moved me. I suppose I had a natural disposition to do things for others, therefore this attracted me. Perhaps it was guilt, I do not know. I do know that I would not have done it on my own and I respected my good friend's opinion.

The concept of volunteering with the aim of making a worthwhile contribution was not the primary driver but rather part of the packaged experience being presented. The complete package is what motivated them to pursue the volunteer tourist experience as can be seen by the following comments:

Chris: The whole experience attracted me: to see another country; to do something worthwhile; to give myself, I liked the idea of helping others.
Hari: I wanted to experience a different culture. I wanted to go with a group of people and have fun. I suppose I also wanted to help others.

In talking about their motivations, respondents commented that they were looking for an alternative type of tourism; to see a country and really experience the culture:

Tama: To see another country but I did not want 'normal' tourism. I had already been to a Pacific Island Resort with my family. I did want to experience another culture and see another country.
Iwi: Yeah, to get first hand experience of the people; alternative tourism I suppose.
Morgan: I suppose the exoticness of India. To not just be a tourist, but to live day to day with the people. To experience the people and the culture, the students, people from the slums, people who set up the schools.

Although not all the participants were motivated by culture alone as the following quote demonstrates:

Aata: The main thing was seeing another country. I think I would have gone no matter where the project was. It was convenient that it was overseas

because I wanted to go overseas. I do not think I was motivated to experience a culture, hey I was young and I was not very 'deep' at the time. The
common thing for most young people in Oz [Australia] at that time was to
go to Bali to check out the beach scene, get drunk, most of us did not have
a lot of interest in culture, it went over our head.

One should perhaps ask if this latter view relating to the perception and role of the
cultural experience is representative of the majority of young people in Western
societies, or at least Australia. If it is, this may be due to the materialism and
hedonism of modern Australian society and the lack of opportunity to reflect on
the non-material dimensions of life such as culture. These accounts highlight that
the volunteers were not seeking or expecting a life-changing experience.

The experimental nature of volunteer tourism

Cohen (1979) identified five modes of the tourist experience, one being the experimental tourist who seeks meaning away from their normal abode. Wearing *et al.*
(2008) argue that Cohen's experimental mode is relevant to the volunteer tourist's
journey into self. The narratives expressed a range of experiences and emotions.
They recalled the difficulties they encountered:

Jae: There were so many different people in the group and some of the hardships in the accommodation were difficult; no hot water, limited utensils to
cook, eating lunch everyday on the back of a truck. This helped me get out
of my western middle class comfort zone and I have tried to stay out of it
ever since.

And what they enjoyed:

Wiri: I liked the social side of the group I was with, meeting and living with a
range of young people. It extended me and I learnt heaps
Tama: The best I thought was not only experiencing but living the everyday life
of the Tongans.

The organised structure of the project allowed some time for shopping, out-
ings to the beach, adventure tourism activities and sightseeing and groups can
be hosted by the more affluent people in the host community or expatriates.
However, the recollection of moments of joy and happiness reported by the
respondents were not associated with the consumption of material goods and
personal indulgence but rather in the act of doing things for others; relating and
being with people.

The narratives described suffering and poverty reflected in a diverse range
of reactions such as tears, grief and escape. Yet upon refection the volunteers
also identified the positive context of suffering and what they could take away
with them:

Toni: The people did not have much but they were happy with what they did have. They were not materialistic, while at home we give so much importance to material possessions and we are not happy.

Aata: The people were friendly and happy even though they had no money and some did not know where their next meal was going to come from. They were the poorest of the poor but they were smiling. I am not being patronising; it taught me a lot on how to cope with hardship.

Iwi: By being confronted with suffering, depravation and poverty I discovered that it is only through embracing suffering in life that you forge your identify as a person.

The volunteers' personal encounters with the host community in their everyday life and through conversations allowed them to reflect on culture and religion and the role these play in peoples lives.

Morgan: The caste system, looking back, I think this is what shocked me. I was on a local bus with kids from a well-to-do school and I asked them what caste they were from. They were from the soldiers/politicians caste and people from a higher caste did not seem to have compassion for the poor. The school girls and university students I met from more affluent environments had never been to the slums, they did not associate with people outside their caste system. Reincarnation is linked to the caste system and the caste system is connected to their religion: Hinduism. Religion colours the culture and the social system.

Aata: Looking back, in hindsight, it helped me appreciate history and culture. I now research before I go to a place. I am also aware of the impact of religion on the culture and it helps me take an interest when I see the history of religion intrinsically linked to the culture. It makes being a visitor/tourist so much more interesting. The Philippines was a catalyst for all this.

Besides the experience opening up respondents' cultural horizons and helping the volunteers to place religion within a cultural framework, some also underwent a religious or spiritual experience. However, religion had not been a part of their life prior to the trip as expressed in the following statements:

Ali: Religion and faith had little bearing on my life at the time.
Hari: Before the service project I was indifferent to religion.

The volunteer tourists came from secularised societies in which religion, if any, is situated in the realm of the private and is generally not manifested in a significant way in the wider aspects of their culture and society. This attitude to religion was in stark contrast to the host communities where they were doing their volunteer work. Religious symbols were in the homes, in the streets, in shops and in vehicles used for public transport. The people talked freely about the role of God and religion in their life and this had reportedly led to the following experiences:

Jae: This exposure made me stop and think about what are the most important things in life and I realised that God was the most important thing.

Morgan: Their simple and deep faith helped me see that I could also share in their faith and strive each day to become better and everything I do can be for the glory of God. Everyday we can put love into what we do and this gives glory to God.

Ali: This whole new world of religion was opened up before me. I saw that these people's simple and trusting approach to life was based on their faith. I compared this to the people around me at home: self-reliant, independent, stressed out, taking on too much, control freaks but having faith in nothing but themselves. I realised what a shaky foundation one has to rely just on self, no wonder people are insecure. I started my quest for a higher power and a religion at that point.

Reflection on their behaviour and attitudes during the volunteer experience led to their journeys of discovery (Lyons and Wearing, 2008) identifying and reassessing their core values and what they wanted to keep and what they wanted to change.

Wiri: We were a big group and it was hard to be selfish. Everyone had to be responsible to contribute and you had to try and get involved. Because I was shy I think I came out of myself a little. The whole experience helped me come out and socialize more and I have made the effort to do this since.

Chris: The people on the service project along with the conversations I had with the local people helped me develop more 'depth' and values I suppose. I now think more deeply on how the world works and not just on the surface. I did not realize it at the time but the service project helped me to grow and travel without my family.

Aata: It brought about a complete change in my attitude. I went because a good friend went. I suppose I was generous but I definitely did not have this disinterested self giving, I was usually seeking myself. I did come back with that, it was what the others in the group and the local people had.

Jae: My approach to life changed. It started by being reflective. I reflected on my superficial life: the binge drinking, the marijuana that was becoming a habit, just seeking a good time. I realised I had no purpose or direction in life, I just existed for myself and my selfish pleasure yet I was not happy. I decided I wanted to change all this. Already during the project I started thinking about others, being a selfish cow was getting me nowhere. I also decided I was going to think a bit more deeply about things and try and reflect on my behaviour.

The long-term life changes

The study sought insight into the potential long-term and lasting impacts of the volunteer tourist experience on the participant. As such, the interviews were conducted at least seven years after the experience and, for one participant, 18 years had transpired. The aim was to demarcate the long-term changes and the experiences felt during the project. However respondents jumped from the past to the present and back to past and hence there is some overlap in the themes identified. It will be argued that this blurring of the past and the present provides further evidence of the deep long-term life-changing impact of the volunteer experience, expressed many years after the actual experience.

Wearing *et al.* (2008: 69) claim 'that the effects of volunteer tourism on one's development of self may indeed be quite profound and carried on into other aspects of one's life'. A prevalent theme in the narratives was the change from 'ego' and self-centredness to 'other' and 'giving to others' through one's relationships, family, work and leisure activities and this has given more meaning to their lives.

Aata: You know that saying 'in giving you receive', I really felt that. I did have a holiday despite all the hardships. When I came back to Oz [Australia] I was really looking forward to going back to work, I suppose to give. Not only to give to others and society through my work but also in my wider social and community relations and especially to my family.

Hari: With the service project you are doing things for others and as a consequence you get a lot more out of it for yourself. It is a paradox isn't it? Discovering this has not only helped me continue to live this it but also helped me pass on the experience, and wisdom I suppose, to others.

Iwi: I came back with a treasure: the realisation that I am happiest when I forget myself and give myself to others. I can honestly say that I have tried to live this out every day of my life these last twelve years since the service project. I can't say I have always succeeded and I have failed and withdrew into self on many occasions but I can say I have tried. I have been richly rewarded but I also know I have passed on this treasure to others.

The narratives reflected the long-term change in their values in relation to materialism and consumerism.

Toni: Yes, I am not so materialistic and I am much more discerning before I buy something.

Tama: The experience helped me become aware of the materialism of our western societies and the aggressive marketing that wants us to consume more. I have made a point to resist this and I have tried to educate others, starting with my children. I still find it liberating to value myself for who I am and not for what I possess or how I appear. My wrinkles are increasing yearly but I do not need botox and I do not care about the scales others use to judge me.

Ali: My life is now full of spiritual values rather than material values and they continually provide more meaning to my life.

The role of family and community became the priority rather than personal goals.

Aata: My attitude to family and children changed. I was a yuppie, a successful professional earning big bucks, a career woman and then I realized
the value of family and children. The service project started all this. I
am now married with a family. I think I have made a better choice in a
husband, similar values. The family rather than career is what is important to me. I also now appreciate the value of education and the unique
role parents have in the education of their children.

Morgan: Many times I have sacrificed my personal plans for my immediate family, extended family and sometimes the communities I am a part of. I
know I would not have made these choices if I had not gone on the
volunteering project.

The narratives highlighted active involvement in advocacy and social justice
issues.

Wiri: It opened up my eyes to social justice issues: wars in the Congo, Aids
in Africa etc. especially how injustices lead to poverty. The Service
project helped me have a social awareness, a dimension that I otherwise would never have had. I have tried to continually develop this and
my husband and I plan to dedicate our 'middle years', once the kids
leave home, to work abroad in less developed countries.

Morgan: One of the hardest things was the orphanage for the disabled kids. You
drove in and you saw manicured gardens, the offices of the staff were
beautiful, it was a grand colonial building. Where the kids were it was
filthy. The kids were sitting on dirty mats, the beds were filthy. The
children were never taken outside they were kept indoors the whole
time. The people spoke to the head guy to take better care of the kids. It
really taught me to look behind things and now I am more vocal about
family and social issues and I have developed quite a reputation.

Chris: It provided me with a deep awareness of poverty and the causes behind
those who have and those who have not. We in the West are just like
the rich people in the industrial revolution oblivious to the poverty and
squalor around them. I have been active in trying to raise this awareness and through my research work I have tried to identify barriers to
sustainable development.

The volunteer tourist experiences also brought about a significant change in the
types of tourism products that were chosen in subsequent years. A common theme
in all the narratives is a complete rejection of any form of mass tourism. The
respondents describe the search for a greater personal engagement with the local
people in the countries they visit:

Aata: Definitely, since then, every trip I have gone on has had a purpose. I
have not looked at overseas trips as a pure holiday or pure tourism. I
went to the Philippines two more times on a service project and New

Zealand on a service project. I have also attended conferences overseas and spent my honeymoon overseas. Even on my honeymoon we were not pure tourists, we did see places but we made a lot of effort to see and be with the local people. The way to know a place is to be with the local people and partake in their everyday life which is not what you see from a hotel or a tour.

Where did you go for your honeymoon? France and Italy. I had made connections over the years and I valued catching up with these people more, seeing them and spending time with them than sightseeing. I will never choose to go on the standard tourist trip again, there is just no go.

Morgan: There was a huge difference between the India trip and the cruise I went on. The cruise was tailored to meet every type of whim, a whole range of entertainment, pure indulgence, 24 hour buffet, everything tailored for you. You were there to get for self but it had a lot less impact on me. The service project was definitely more fulfilling; sharing yourself, giving your time and talents, the cruise was taking. In the case of the cruise I was with my husband's family and I was giving myself to the family but otherwise you felt 'blah' on the cruise.

Jae: Now with a young family I see the need to go away and relax but I will definitely encourage my children to experience what I experienced, I think it is really important for their learning and their development. This is a great way for them to experience the culture, the people, to give themselves and do something useful.

Hari: I will never go to another country and stay in a normal hotel as a normal tourist. To me it is a waste of time and money and the only one benefiting would be me. On second thoughts I would be bored and miserable.

Conclusion

This chapter has provided evidence of the transformative potential of volunteer tourism (Wearing *et al.*, 2008) and the 'long-lasting reworkings of the "I"' (p.69). Wearing (2001) also claimed that the volunteer tourism causes 'value change and changed consciousness in the individual that will subsequently influence their lifestyle' (p. x). The chapter supports this claim in providing evidence from a longitudinal study. The narratives describe a resistance to a materialistic and consumer society, a sustained consciousness of one's role within the family and society, examples of advocacy and a commitment to social development and a rejection of mass tourism.

These experiences were not the 'normal' tourism experience and were life changing. The experiences were described as: 'mind blowing'; emotional; 'I suffered'; difficult; rewarding and spiritual. They led to the discovery of self: 'I need to forget about myself'; 'it helped me to reflect'; and raised fundamental questions such as: How can I go on living my life as I used to? What are my values? And how should I choose to live my life? This self-discovery led to the further discovery of their role in society, embracing the notion of 'gift of self' in their professional careers, family life, friendships and community involvement. It led to the conviction of their role in combating poverty, fostering justice and equity in society and the importance of solidarity with their fellow human beings rather than self-seeking individualism. They reflected on their values, the way materialism and consumerism dominated their outlook on life and the decisions they made. The volunteer tourism experience made a significant impact on them at the time and their future choices and direction in life.

References

Arai, S. M. (2000) Typology of volunteers for a changing sociopolitical context: the impact on social capital, citizenship and civil society. *Society and Leisure*, 23(2): 327–352.

Atkinson, R. (2002) The life story interview. In J. Gubruim and J. Holstein (eds), *Handbook of Interview Research: Context and method*. Thousand Oaks, California: Sage.

Beverly, J. (2000) Testimonio, subalternity and narrative authority. In N. K. Denzin and Y. S. Lincoln (eds) *Handbook of Qualitative Research* (2nd edn). Thousand Oaks, California: Sage, pp. 555–566.

Broad, S. (2003) Living the Thai life – A case study of volunteer tourism at the Gibbon Rehabilitation Project, Thailand. *Tourism Recreation Research*, 28(2): 63–72.

Callanan, M., and Thomas, S. (2005) Volunteer tourism. In M. Noveli (ed.) *Niche Tourism*. Oxford: Butterworth-Heinemann, pp. 183–200.

Carlson, J. (1991) *Study Abroad: The Experiences of American Undergraduates in Western Europe and the United States*. New York: Council of International Educational Exchange.

Chase, S. (2005) Narrative inquiry: multiple lenses, approaches and voices. In N. K. Denzin and Y. S. Lincoln (eds) *The Sage Handbook of Qualitative Research* (3rd edn). Thousand Oaks: Sage, pp. 651–679.

Cohen, E. (1979) A phenomenology of the tourist experience. *Sociology*, 13: 179–201.

Gubruim, J., and Holstein, J. (2002) *Handbook of Interview Research: Context and Method*. Thousand Oaks, CA: Sage.

Lieblich, A., Tuval-Maschiach, R. and Zilber, T. (1998) *Narrative Research: Reading, Analysis and Interpretation*. Thousand Oaks, CA: Sage.

Lyons, K. D. and Wearing, S. (2008) *Journeys of discovery in volunteer tourism*. Wallingford: CABI Publishing.

McGehee, N. G. and Santos, C. A. (2005) Social change, discourse and volunteer tourism. *Annals of Tourism Research*, 32(2): 760–779.

Ryan, C. (1997) *The Tourist Experience: A New Introduction*. London: Cassell.

Simpson, K. (2004) Doing development: the gap year. Volunteer tourists and a popular practice of development. *Journal of International Development*, 16: 681–692.

Stebbins, R. A. (2004) *Volunteering as Leisure/Leisure as Volunteering: An International Assessment*. Cambridge, MA: CABI.

Stoddart, H. and Rogerson, C. M. (2004). Volunteer tourism: the case for Habitat for Humanity South Africa. *GeoJournal*, 60(3): 311–318.

Wearing, S. (2001) *Volunteer Tourism: Experiences that Make a Difference*. Wallingford: CABI Publishing.

Wearing, S. (2002) Re-centering self in volunteer tourism. In G. M. Dann (ed.) *The Tourist as a Metaphor of the Social World*. Wallingford: CABI Publishing., pp. 239–262.

Wearing, S., Deville, A. and Lyons, K. D. (2008) The volunteer's journey through leisure into the self. In S. Wearing and K. D. Lyons (eds.) *Journeys of Discovery in Volunteer Tourism*. Wallingford: CABI Publishing, pp. 63–71.

Weinmann, S. (1983) *Cultural Encounters of the Stimulating Kind: Personal Development through Culture Shock*. Michigan: Technological University.

Zahra, A. and McIntosh, A. J. (2007) Volunteer tourism: evidence of cathartic tourist experiences. *Tourism Recreation Research*, 32(1): 115–119.

8 Self and society in voluntourism

A thirty-year retrospective analysis of
post-trip self-development of volunteer
tourists to the Israeli kibbutz

David Mittelberg and Michal Palgi

Volunteer tourism

Volunteer tourism as a concept derives from the wider concept of 'alternative tourism' (Wearing, 2001). It is a unique form of tourist behavior driven by an ideology. Its prime motivation is to *contribute* to society, and/or to a given community, by moving and living in it on a temporary basis, working as a volunteer to improve the quality of life of that community. Thus, the pure tourist experience becomes only a secondary motivation. Nevertheless, this secondary motivation plays an important role in constructing the overall travel experience. Hence, the intensive host-guest interactions, the exposure to local cultures, the cross-cultural experiences and the local tourist attractions are all perceived by volunteer tourists as a major benefit alongside the fulfillment of their ideological urge to volunteer (Mittelberg, 1988; Wearing, 2001).

To a great extent, volunteer tourism is a form of travel which allows the individual to pursue two goals. One is the urge to escape from one's social environment and experience different cultures with different norm and values systems. This form of escapism varies from the regular motivation to flee from routine and boredom. The second is the need to enrich self-identity by broadening one's perspective on cultures and societies (Wearing, 2001). Although there is still a major debate amongst researchers as to what extent these types of motivations are too generalized and thus, perhaps, irrelevant, they can still be used to illustrate the sociological drive behind volunteer tourists (MacCannell, 1992; Mansfeld, 1992).

Types of volunteer tourism and the prime ideological motivation vary. The typology could easily be explained by the 'push' and 'pull' factors that shape the decision to undertake such travel. While the motivational 'push' factors identified above have been well documented in the literature, the 'pull' factors have been mostly ignored, especially in the tourism academic literature. Mittelberg (1988; 1999) in his work on volunteers to Israeli kibbutzim defined the 'pull' factors as inviting economic, socio-cultural and political settings that call for assistance or help. Thus, in most cases, the host community seeks low-cost manual workers to solve its workforce shortage and finds volunteering an appropriate solution. Hence, such volunteers may be treated as inexpensive labor rather than mere international tourists in pursuit of fulfilling their travel motivations. But what of these motivations themselves? Mittelberg (1998) has shown that both guests and hosts bring a plurality of motives to the encounter between them (from what Cohen (1979) calls 'touristic' recreational motives, to meaning centered existential ones, so that the character of the resultant *experience* for the volunteer is an outcome of the negotiated resolution of the congruence or lack of it between these pluralities of motive (Mittelberg (1988: 63–83). Pearce and Coghlan (2008) citing Pearce (2005) report that they were able to distill from 14 recurring themes of motivation in tourism research, the three core motivations of (1) the search for novelty, (2) desire for escape and relaxation (3) opportunities to build relationships (Pearce and Coghlan, 2008: 140). Interestingly they determine that travel veterans 'emphasize involvement with host communities and settings as important to them' (2008: 140).

Wearing (2001) also reports that while much research has been done on what he calls the vocabularies of motive of volunteer tourists, little research has been completed on the impact of these experiences on the development of self. It is precisely this theoretical question this chapter will address, in an examination of a case of volunteer tourism that both predates and is largely unacknowledged in the contemporary literature on voluntourism. Following Wearing's pioneering discussion of the impact of the volunteer experience on the development of self especially amongst young adult volunteers, this chapter describes the high degree of self-development reported by kibbutz volunteers while exploring the mechanism by which this self-development takes place utilizing the concept of existential authenticity coined by Wang (1999) and utilized by Lev Ari and Mittelberg (2008).

Volunteers to kibbutz

A kibbutz volunteer is an international tourist on a working holiday who contributes his/her labor in the kibbutz in literal exchange for free board and lodging, some small pocket money and access to the collective facilities of consumption, dining room, leisure amenities and so on, shared by all kibbutz members, In fact they are temporary working guests in the kibbutz. National benchmark survey data collected by one author in 1979 estimates that in the 1970s up to 30,000 volunteer tourists visited the kibbutzim per annum, while the total adult population of the kibbutzim totaled only 120,000. This would make it clear that volunteer tourism was very much a hallmark of a significant section of young 'alternative' oriented baby boomers and not necessarily a phenomenon beginning with Gen X or Y as cited by Pearce and Coghlan (2008: 133–134). Pearce and Coghlan call for an understanding of the dynamics of volunteer tourism, particularly that volunteer tourism is 'beyond simple altruistic motives' but is 'a complex process where the individual is at the core of understanding the volunteer tourism experience' (2008: 142). Indeed they point to the role of tourism, especially cross-cultural tourism, in shaping issues of identity, and human growth is an insight from the focus on volunteer tourism. As our data will amply show, volunteers to the kibbutz from the late 1970s through to the 1990s overwhelmingly report the consequence of personal growth as a result of their visit and so it was also of volunteers in the 1980s (see Mittelberg, 1988: 117–126).

Pearce and Coghlan (2008: 134) offer a useful definition of volunteer tourism activities 'as being performed as a service for others in a novel setting for which an individual pays' (see also Wearing, 2001). The evaluation of this experience according to Pearce and Coghlan rests on four dimensions that determine a good volunteer experience: 'skill development and new personal insights, a good social life, experiencing novelty and contributing to a worthwhile project' (2008: 136).

But how are we to understand the dynamics of this experience that lead to a good experience or otherwise? Pearce and Coghlan (2008) utilize fruitfully equity theory. This theory requires the actor and the researcher to identify inputs and outputs as perceived by the actor and to determine the stress arising from any imbalance between them, which determines the relative weight of what the volunteer contributes, compared to his perception of what he receives. They indicate how this might impact the individual.

Mittelberg (1988) has referred to the same dynamic through a different theoretical framework in order to see how this same stress or incongruence may lead to different forms of guest host institutionalization. The volunteer traveler has recourse to what Cohen (1974) aptly calls the 'environmental bubble'. The environmental bubble serves both to preserve some familiarity within a strange environment and to familiarize the traveler with strangeness. Different types of travelers have different bubbles varying in their degrees of cohesion and penetrability, meaning the degree to which the actor is prepared to leave the environmental bubble, in addition to the degree in which the traveler's local role is institutionalized. The problem here is what are the 'limits of incongruence' between the worldviews of strangers and hosts that can be mutually tolerated. By way of example, the kibbutz volunteer bubble organized by the hosts serves both to bridge the gap of task orientation to volunteering work, but also to the unique values of kibbutz society in daily life. The kibbutz volunteer may have all the appropriate ideological characteristics and thereby be, in principle, a candidate for membership. On the other hand, the volunteer requires the environmental bubble, just as much as the hosts insist on it, primarily because of the host's unwillingness ordinarily to consider some as potential candidates for membership on national, ethnic or socialist grounds.

Most pertinent to this discussion of reciprocity is Cohen's (1974) discussion of the effects of temporariness or permanency on the accommodation process itself. The host will be prepared to tolerate a greater deal of cognitive and normative autonomy of the stranger the less permanent his or her stay, since the more permanent stay brings with such autonomy a concomitant *threat* to the worldview of the hosts.

At the same time, the shorter stay implies an increasing need for an environmental bubble to assuage the threats of disorder felt by both the hosts and guests, as the kibbutz can tolerate neither temporariness nor nonconformity indefinitely, as a general rule. Thus the environmental bubble serves not only the strangers but also to protect the hosts from the long-term effects of the continuing presence of strangeness.

The sociological dynamics of the guest–host encounter

The sociological dynamics of the guest–host encounter as such was first ana-lyzed by Sutton (1967), who considered which aspects of the structure of that encounter contributed to the cross-cultural understanding and which did not. Put another way, Sutton (1967) recognized that there were some factors in the structure of the encounter itself which lead to 'misinterpretation of motives and misunderstanding', while others lead to 'smooth interaction and an enhance-ment of mutual understanding' (p.221). Neither outcome is then a necessary outcome of the cross-cultural encounter, despite the cultural distance between guests and hosts.

Sutton (1967) then identified five social characteristics of the tourist–host encounter, which, in his view, together point to the different possible outcomes. The main parts of each will be dealt with briefly, in turn. The first is the mutual recognition by both guest and host of the 'transitory and nonrepetitive character' of the relationship between them. Now this temporariness can lead to mutual tol-erance, as Cohen (1974) has asserted above, but it can also in Sutton's view 'make each more predatory and exploitative of the other precisely because an extended relationship is not anticipated by either. The second characteristic that follows from the first, is that both guest and host share 'an orientation for immediate gratification'(1967: 221): the guest, in order to maximize the experiences and the host, to maximize the price that can be charged for providing the services.

The third characteristic lies in the inherent cognitive asymmetry between guests and host. The host knows what the guest needs and does not need to know, in order to fulfill the goals. This dependency can lead to satisfactory social exchange or, alternatively, increase the power of exploitation of the host with the resultant resentment of guest toward the host. The fourth characteristic relates to the fact that on the one hand, the guests can bask in the euphoria of 'new' or non-routine experiences, while on the other, the steady stream of new experiences can be stimulating for the host.

The last characteristic is the problem of 'the relative congruence of the two par-ties' norms and values' (Sutton 1967: 222), which, especially in the cross-cultural situation, can well lead to 'friction and misunderstanding' as the degree of incon-sistency between the two cultures increases. Basically, on all these dimensions, while the parties are engaged in an exchange relationship based on tolerance and generosity, travel can increase understanding. However, drawing on the antitheti-cal themes of these same dimensions, Sutton observes that

> when the transitory and non-repetitive character of the encounter is combined with the desire for immediate gratification, a considerable lack of restraint may afflict both host and visitor … the probability that both host and visitor will impute selfish purpose to the other and set the stage for a degenerating cycle of misinterpretation is great indeed.
>
> (Sutton 1967: 223)

Volunteers in the kibbutz: problems in the guest–host encounter

What, then, are the major problems in the guest–host encounter? First and fore-most is the very strangeness of the new social world into which the guests have temporarily entered. This encounter, occurring as it does at the cross-cultural interface, requires its participants to cope with a new language, customs, norms, etc., indeed the 'rules of the game' of being a guest in kibbutz. Typically, the guest has to generate simultaneously new interpersonal relationships both with fellow guests who *share* this strangeness as well as with the hosts.

Thus, the guest needs to cope with the cognitive questions of communication or language, as well as the adaptation to a way of life different from home both in its structure and content. In order to amplify this discussion, different spheres of the guest–host encounter will now be considered.

The volunteer workplace

There are a number of potentially problematic aspects to the guest–host encounter in the workplace: first, the fact of having to work where you are assigned, with only a limited choice of work place; second, the type of work, which is often manual or unskilled and menial and in some cases, in the light of the guests' home values, can even be seen as demeaning; third, the relative permanency in an unde-sirable workplace, in which kibbutz members work, if at all, only *temporarily*; fourth, relatively little contact with hosts in the workplace, while the contact that exists is often of a formal supervisory nature. Thus, the problem for the guests is to get a more or less permanent work placement in a job that offers some sort of intrinsic satisfaction, if possible with avenues for positive guest–host interaction.

Leisure: after-work hours

Here the volunteer encounters a number of problems at different levels. At the superficial level, the guest needs to negotiate access to and utilization of the pub-lic facilities, e.g. the dining room, laundry, supermarket, etc., as well as the public leisure amenities generally at the disposal of the guest, for example the members' coffee room *(moadon),* discotheque and sports facilities (pool, tennis, etc.). This is generally relatively easy to accomplish. At the deeper level, the major problem is the actual generation of any guest–host encounter at all, which, as we will see, is held in high esteem by the guests.

This problem is an outcome of the relative impermeability of the institution-alized environmental bubble in which the guests live. In fact, many volunteers are allotted kibbutz families to visit, but this is a potential encounter that needs itself to be realized and institutionalized successfully in practice. Moreover, the guests will try, in so far as it is considered to be important to them, to generate additional means of guest–host encounter in different social settings. Here the differential permeability of the bubble on both demographic and normative lines becomes revealed. Indeed, guests, especially women, may themselves become the objects of 'reverse tourism' of the hosts, which they may be able to turn to their own advantage in their efforts at social acceptance, though this is a matter of negotiation (Mittelberg, 1988: 104). It was also found that much of the social activity in this sphere of their life is carried on within the environmental bubble, so that guests can either accommodate themselves to this state of affairs or devise strategies for crossing over to the host-world (Mittelberg, 1988: 85–116).

While Sutton (1967) grants the possibility that the encounter can lead to greater understanding between guest and host, he quite insightfully pointed out that when the encounter is characterized by transitoriness, malimputation of motive, and the mutual desire for immediate gratification, then it may lead to a degenerating cycle of misinterpretation.

Yet why is it that despite the lack of symmetry between the imputed purpose of the encounter of guest and host on the kibbutz, this degenerating cycle does not develop? This is so due to the institutionalization of the encounter where we find the practice of family adoption of guests, which serves as a possible basis of social exchange. But in terms of the intergroup relations between guests and host society, adoption actually provides the 'idiom' of interpersonal social exchange, which masks the economic exchange that is actually negotiated between the volunteer sector and kibbutz society.

The structure of the 'bubble'

The kibbutz volunteer guests, therefore, live in their own ecological tourist space with its formally defined obligations, predetermined number of organized excursions, fixed rate of pocket money, unlimited access to kibbutz recreational facilities, etc., in return for which the guests work a *shorter* day than do the members, without most of the additional burdens of after-hours roster work, that members themselves are obliged to do. Finally, of course, the volunteer sector is served by a kibbutz member who acts as coordinator or mediator between the sides.

At the same time, the guests are freed of those distinctive normative obligations that hold for kibbutz members vis-à-vis the hosts, and vice versa. Thus, the kibbutz discharges its debt to the guests as a group, neutralizing the potential power that the group might have (were it to exercise collective power in the form of withholding services) by negotiating the terms of intergroup economic exchange in the clothing of interpersonal social exchange.

Pearce, echoing some of these concerns, has argued that understanding volunteer tourism 'requires that the interacting parties be viewed as actively constructing their experiences and their relationships' (2005: 134). In both views, the volunteer needs to resolve imbalance by either heightening values of input or lowering their price (Pearce, 2005 in Pearce and Coghlan, 2008: 137).

Authenticity in volunteer tourism

Wang (1999) offers an important discussion on the nature and role of authenticity in tourism. After surveying the literature and debate surrounding the differences between objectivist authenticity that derives from the authenticity of originals, to constructed authenticity that derives from the projection of symbolic meaning on to the toured objects, Wang (1999) suggests a third type of authenticity that he labels existential authenticity. Existential authenticity consists of 'an existential state of being that is to be activated by tourist *activities*' (Wang, 1999: 352, author's emphasis). This state of being refers to the tourists' pursuit of their true self. For Wang (1999), the loss of self is identified as part of the modern condition.

Wang further distinguishes between intra-personal and inter-personal existential authenticity. The intra personal, which refers to bodily feelings, is reserved for the making of self in the transcendence of the routine, perhaps on a beach or through adventure (1999: 361–362). The inter personal refers to the quest for an 'emotional community', 'tourists are not merely searching for the authenticity of the *Other*. They also search for the authenticity of, and between, *themselves.*' (Wang, 1999: 364, emphasis as in the original). The communitas of the touristic group serves as a source of authenticity of the members and it is their social construction.

Methodology

The process

At the beginning of 2005 we heard from the volunteers department in the kibbutz movement that veteran volunteers were planning a get together in the summer. We decided it was a good occasion for getting some insights into the aftermath of their kibbutz experience. This was especially important for us since we had several studies on volunteers while they were in the kibbutz and very few (Mittelberg, 1988) on their perceptions of its after effects. There are, however, some studies (Palgi *et al.*, 1995) that show that the socio-cultural adaptation of new immigrants to life in Israel after a six-month stay in the kibbutz was better than that of those who did not. We got in touch with the organizers who are the owners of the Lostamigos website and asked them to publish on their website the questionnaire. They agreed and the following appears in the Lostamigos website (http://www. lostamigos.net/):

> **KIBBUTZ VOLUNTEER?** The Institute for Research and Study of the Kibbutz and the Cooperative Idea needs your help. Please download and complete this questionnaire …

The sample

The sample was not a representative sample and consisted only of those who knew of the Lostamigos website (an indication of this is that we get letters that arrive almost daily with more questionnaires of those who have just heard about it) and of those who were interested and motivated enough to respond. We therefore should review the data with due caution. In the period 2005–2008 we received 373 completed questionnaires from volunteers who had served in 151 out of the total 267 kibbutz communities. Thirty-three percent of these respondents had volunteered to a kibbutz in the 1970s, 35 per cent in 1980s, 30 per cent in 1990s and the remainder (2 per cent) in the last decade. Sixty-five per cent of the respondents volunteered for less than half a year in the kibbutz, 27 per cent stayed between 6 months and 12 months, while just 8 per cent extended their stay beyond a year.

While this sample is in no way to be regarded as representative of the entire population described above, it does offer an important and rare insight into some of the long-term impacts that the volunteer tourist experience can have on the trajectory of the lives of the young people who engage in it.

Demographic characteristics of the respondents

Forty-nine per cent are male and 51 per cent female; 66 per cent are married or have life partners, 22 per cent never married and 12 per cent are divorced, separated or widowed; 52 per cent were born before 1962 and 48 per cent after; 57 per cent came from English-speaking countries; and 60 per cent came from Western European countries: very similar source countries to those reported by Pearce and Coghlan (2008).

The questionnaire

There were two parts to the questionnaire. The first was quantitative and had socio-demographic questions like age, marital status, profession; informational questions on the volunteer experience, such as in which kibbutz they volunteered, length of stay and number of times they were volunteers; evaluation questions on the impact of the kibbutz experience on their cognition and development of self. The second part of the questionnaire was open and gave the respondents an opportunity to express their thought and feelings in their own words.

The analysis

The quantitative data were analyzed through frequencies, cross tabulation etc. The qualitative part of the questionnaire was analyzed by at least two of the researchers in order to increase the reliability of our interpretations. The analysis itself was done in three steps. First, with the aid of our graduate research assistant, we read and re-read each response and open-coded it, then we did a cross-case analysis and finally we went back to the individual responses to find the way they relate to specific terms and ideas mentioned in the theoretical literature (such as development of self). To validate the plausibility of our categories, we engaged in peer debriefing whereby we, the authors, consulted with one another, allowing us to uncover patterns and emerging themes within the data set.

Findings and analysis

In our analysis below, we have explored how the volunteers evaluate the outcomes of their own kibbutz volunteer experience.

Table 8.1 is divided into four different outcomes of volunteering as perceived by the volunteers themselves: personal growth (a); skill development (b, c); new political insights (d, e); social capital-ties with Israelis (f) and with other volunteer peers (g).

Personal growth

As seen in the Table 8.1, 84 per cent of the volunteers state that their volunteering activity resulted in personal growth. Of those who stayed over a year in the kibbutz, 92 per cent felt so while among those who were volunteers for 6 or less months 77 per cent reported the same. In addition, 70 per cent stated that volunteering impacted to a very large extent their life and the paths they chose after their stay. Interestingly, there were no age differences in the responses to these questions.

Skill development

We looked into two types of skills out of those that might have been achieved during their stay on the kibbutz: first, the learning of Hebrew; second, learning about Israeli society. Both were chosen because they were seen as facilitators in the understanding of the local culture. Twenty-nine per cent of the volunteers learnt Hebrew during their stay and 61 per cent learnt about the Israeli society. Forty-two per cent of the volunteers who were born before 1957 learnt Hebrew while only 20 per cent of those born after 1968 did so. Acquiring Hebrew was linked to duration of stay, so that 48 per cent among those who stayed for a year or more in the kibbutz, versus 17 per cent of those who stayed half a year or less, reported that they had learnt Hebrew.

Table 8.1 To what degree did your participation in the volunteer program bring with it the following consequences? (Percentage)

		To a small degree	To some degree	To a large degree	Total
a.	Personal growth	2	14	84	100 (314)
b.	The learning of Hebrew	51	20	29	100 (308)
c.	The learning about Israeli society	8	31	61	100 (310)
d.	Understanding Israel	2	19	79	100 (313)
e.	A better understanding of Middle East conflict	13	29	58	100 (312)
f.	Establishing long-lasting ties with Israelis	42	19	39	100 (310)
g.	Establishing long-lasting ties with other volunteers	37	25	38	100 (312)

New political insights

In this category we included understanding Israel and a better understanding of Middle East Conflict. Seventy-nine per cent of the volunteers reported that they could understand Israel better after staying in the kibbutz. A larger percentage of the older volunteers related that as a result of their stay in the kibbutz they understood Israel better. Thus, 83 per cent of those born before 1967 understood Israel better compared to 70 per cent of those who were born after 1967. Also, the longer the duration of the stay, the better they understood.

Fifty-eight per cent of respondents stated that after volunteering in the kibbutz they had a better understanding of Middle East conflict. No age differences were found relating to this issue but those who stayed longer in the kibbutz understood the Middle East conflict better.

Social capital

Two forms of social capital were reviewed: first, the establishment of long-lasting ties with Israelis, second with other volunteer peers. Similar percentages of volunteers (38 per cent and 39 per cent) stated that they had maintained those long-lasting relationships to a great extent; 54 per cent of those who stayed over a year in the kibbutz had long lasting ties with the kibbutz compared to 34 per cent of those who stayed a year or less. A larger percentage (54 per cent of those born after 1968) among the younger volunteers had stronger ties with other volunteers than the older (21 per cent). Also, among those that were on the kibbutz before 1985, only 27 per cent reported having long-lasting ties with other volunteers while among those who came since 1985, 51 per cent reported having such ties. Perhaps this could be a result of the longer time that has passed since their volunteering period or rather the development of internet and email only in the later years under review.

Participant satisfaction from the experience

We wanted also to know how, in retrospect, is the volunteer experience and the Israel experience rated; a high, 84 per cent of the volunteers rated the kibbutz volunteering experience as excellent and 77 per cent the overall Israel experience. In addition, 59 per cent stated that they were still extremely attached emotionally to the kibbutz.

The volunteer kibbutz experience: a retrospective analysis

Overall, the responses were predominated by positive reports of their volunteer period on the kibbutz. All the respondents listed on the following pages are identified by their gender and the year of their first volunteer period on kibbutz. For most of the volunteers their first visit in Israel was in their 20s, often it was their first and the longest stay outside their parents' home. They described it as an amazing, incredible and unique period.

> Volunteering in a kibbutz was the single greatest experience of my life, and I had traveled extensively prior to my experience. I had driven around the U.S. presenting multi-media high school program right after college and I had spent a semester living in San Franciscio, but nothing compared to immersing myself in kibbutz culture.
>
> (Male, 1988)

> I am an old timer volunteer, having first arrived in Israel in 1973 (just before the Yom Kippur War). I was 16 years old. I would like to say that this was the best experience of my life and one that never left my psyche. Kibbutz life, then, was the best. There was nothing bad about it. The people were great to work with (mostly!), the sense of community was great, the friendships between kibbutzniks and other volunteers were fantastic also.
>
> (Female, 1973)

The volunteers valued the chance to meet young people from different countries, cultures, religions and different walks of life. They also identified the period on the kibbutz as a formative and enriching period resulting in self-development. Some declared that the time they spent on the kibbutz/Israel changed their life forever. They had grown up, learned a lot about themselves and the other; they developed tolerance, self-confidence, self-esteem and personal awareness. Some volunteers reported that the experience of a different framework of life such as on the kibbutz led to a change in their career path. It is noteworthy that the volunteers related all of these changes to the unique volunteering experience claiming that it could not have happened in another framework.

> The kibbutz taught me to be frugal, to be community minded, to slow down, to be respectful, to be tolerant, to be a worker, to be quiet, to listen, to appreciate differences, to appreciate wealth, to appreciate all that life has to offer. Without my kibbutz experience, I never would have moved to Israel, I never would have found the inner peace I have most certainly, would have never become the man I am today.
>
> (Male, 1969)

> I enjoyed the idea of communal working and after Israel I spent my entire working life in public service.
>
> (Male, 1969)

> I think as a consequence of mixing with Israelis and people of many different nationalities, our minds are opened up to more beliefs and we more easily accept people from other cultures and beliefs.
>
> (Male, 1985)

I went to Israel as a 21 year old without any true knowledge of life and came back having met an extremely proud people and gained a healthy respect for the Israeli people and the kibbutz way of life it was one of the most pleasurable and educational experiences in my life.

(Male, 1985)

Some described the time they spent in Israel as an enriching period. Most of the volunteers claim that the volunteering was a life-changing experience. A few claimed it had changed and shaped their life and career path after their period of time on the kibbutz. They became more tolerant people, more accepting of difference; they have learned a lot about the other. The experiences led them to change their opinion and reconsider their point of view.

I felt the Kibbutz period of my life had a profound and lasting effect on me, influencing my values and ethics permanently. It gave me a place and space to consider what life was all about … The kibbutz allowed me to consider different values.

(Male, 1975)

Expectations and frustrations

Before coming to Israel some of the volunteers imagined what the framework of the kibbutz would look like. Some recalled that they were surprised and disappointed to see that the picture they had imagined didn't fit to the reality.

I guess that I had expected the kibbutz to represent some kind of ideal socialist community, which concentrated on communal care, work and engagement. I experienced a very different kind of social engagement at the kibbutz: kibbutzniks were eager to define their individual boundaries or emphasize family boundaries, celebrating the Shabbat in their houses and not in the dining room, and collecting their own private food supply. In this way, the kibbutz community was very much like my childhood experiences of middle-class village-life in rural Denmark.

(Male, 1994)

While negative experiences were relatively few they need to be noted here. One participant felt that the work environment was not pleasant. Some reported that the conditions and the treatment changed from one kibbutz to another. Furthermore some volunteers claimed that they were treated differently from other volunteers and members, because of their religion. Some even felt exploited.

The organization that brought me to this horrible Kibbutz lied about everything. I soon found out that the Kibbutzniks (volunteers-coordinator, workplace, etc) were not very different. It was like a labour camp. The bar was not for volunteers, the pool had been drained, the factory-bonus was not given (financial bonus for mindnumbing factory work) as was promised, there were no trips organized out of the kibbutz, the housing was almost too poor to believe: rotting and humid shacks, volunteers were not allowed to hang out with kibbutzniks (they called us morally corrupt there).

(Male, 1997)

Since I've been to two kibbutzim I have realized that they are very different. At the first one I was, at least after a while of showing interest, very welcomed by the kibbutzniks and I got to know lots of young people and their families. They taught me Hebrew, invited me to their homes, took me on trips etc. At the second one, my experience was not the same. The atmosphere was much rougher and volunteers were mainly working staff.

(Female, 1998)

Developing the self through the volunteer experience

The volunteering experience often built self-confidence and a sense of maturation.

At a time when I was young, seeking, and free to broaden my horizons via travel, Israel was an unplanned treasure and one of the most memorable experiences of my life. I met many people from all over the world as a volunteer, made close friends with kibbutzniks, and learned of the complexities of a country and people I developed a great respect for – hard-working, intelligent, open, respectful, and dedicated to a unique way of life.

(Female, 1972)

I cannot emphasize too much as to the positive effect volunteering on a kibbutz had on my life. I left for 6 weeks in Israel, a shy and sheltered 19 year old from a very conservative background. I grew up in Israel, learned to make friends and how to get along with people, gained self confidence and a lifetime interest in world affairs. For the first time I felt accepted and respected based on my intelligence and wit rather than how I looked or how athletic I was (or in my case wasn't!).

(Male, 1974)

Social relationships

It seems that most of the volunteers made many kinds of relationships with members and other volunteers, from different countries and religions. The kinds of relationships that were reported included marriage, foster family and other simple relationships. Some of the friendships did not last long, only for the period time on the kibbutz, some lasted longer, until today. Although most of the relationships have been between the volunteers, there were only few with the members. It seems like all of the relationships between the volunteers and the members were with young members, members from the opposite sex and with their colleagues. Respondents report that the reason for the lack in the relationships, especially with the older members, was because of the language differences. It was difficult for them to speak in a foreign language. Some of the volunteers had a foster family.

> At the beginning only a small group at the Kibbtuz X knew that I am non-Jewish German. I knew from the invitation that the Kibbutz had to take a tough decision to invite a German person to the Kibbtuz (I was the first non-Jewish German guest in a community after the Second World War, where 80 per cent loved to use their native language). But shortly later I was accepted, because I had a good host family. Two years later, they had already several German backpackers as Volunteers and nobody was challenging them.
>
> (Male, 1963)

Friendships with other volunteers lasted longest; they still get together today. A few met their life partners on the kibbutz.

> I made great and intimate friendships with many other volunteers that I will never forget. We kept in touch for a number of years afterwards and visited each other. Well, after 22 years that has passed of course but I still remember all of them so clearly. I guess we'll all be forever young.
>
> (Female, 1985)

> I still keep in close contact with three of my fellow volunteers (two British and one German) and have attended the wedding of another. Despite the fact that I see these people only once every two years or so I feel closer to them than to friends I see weekly.
>
> (Male, 1988)

Some of the volunteers wanted to connect with young kibbutz members, but felt rebuffed.

> One of the things I found strange at first was the almost complete lack of inter-action between volunteers and kibbutzim. I don't know what I had imagined, but it sure wasn't sitting around the volunteer quarters drinking beer with a bunch of drunken Finns and rowdy Aussies. I quickly got used to it, but I recall how surprised I was at how wary and disinterested they were in us.
>
> (Female, 1985)

Long-term affinity for Israel and willingness to return

It seems like a lot of the volunteers enjoyed their stay on the kibbutz and in Israel; as a result some wish to return back for a visit and others as immigrants. Indeed, a few even expressed the wish that their children and their friends will travel to Israel and volunteer on the kibbutz, irrespective of respondents religion. Not everyone fulfilled this wish: some have not returned because of commitments to their family or to their career, others had financial problems and the most common reason for not coming was the political situation in Israel.

> Unfortunately I haven't managed to return to Israel yet, mainly due to the tense political situation during the past years and last but not least my tight financials as a student. But I'll come back at least once more for sure.
>
> (Female, 1999)

> I would like to take my two sons (ages eight and four) to Israel someday, but the current security situation does not make it prudent.
>
> (Male, 1988)

The volunteers continue to identify with the kibbutz and remain concerned about the situation in Israel. Many want to visit again with their offspring in order to share the experience.

> The welcome and kindness that we were shown by Kibbutniks and others was inspirational. I made such good friends with volunteers and members of the kibbutz that I still have contact with now ... Israel is such a beautiful country it makes me so sad that it is constantly under attack and I worry for the people I have left behind.
>
> (Female, 1989)

> I am now the mother of two small boys aged four and 1 and I can only hope that there will still be the opportunity for them to visit and stay on a kibbutz when they are older. We will definitely be visiting Israel one day.
>
> (Female, 1985)

> That was almost 30 years ago, but people say about me sometimes that I have 'kibbutz manners', meaning that I'm open minded, my door's always open and I raised my children to be generous, curious and inventive.
>
> (Male, 1976)

Some participants report that they learned about the unique culture of the Jewish people and fell in love, as a result some immigrated to Israel and even converted to Judasim. Most of them learned a lot about the Israeli walks of life and culture, local politics, Hebrew and about the Palestine conflict. They had access to the history of Israel, Jewish religion and to the Zionist movement.

> We, the volunteers, were no longer long-distant observers of a faraway conflict played out on television news. We were now part of that conflict, at times in personal danger and constantly reminded that in our part of the world politics was often stripped to the bare essentials of killing and surviving.
>
> (Male, 1988)

I learned a little Hebrew and learned lots about the country and culture. I feel strong ties to Israel although I am not Jewish. I feel strongly about the conflict and Israel's politics. I do hope to visit again especially as my kibbutz (adopting) mother is getting older.

(Female, 1983)

The volunteers claimed that the contribution of the kibbutz volunteer movement for promoting empathy for Israel around the world is much bigger and more effective than the influence of thousands of Israeli Ambassadors. The volunteers identified were concerned about the situation in Israel.

The Kibbutz volunteer movement does for Israel, what a thousand Israeli Ambassadors could never do, not for the lack of trying by the Ambassadors. The Kibbutz movement provides an honest reflection of life in Israel, its people and its traditions. An opportunity to learn of its history, its problems, and hopes for the future.

(Male, 1983)

Kibbutz and its unique framework life

As we reported above the volunteers who experienced the kibbutz way of life before the recent changes regarded it as both interesting and enriching.

Utopia it was not, I certainly realized that early on, but people worked together, all on an equal basis. I loved it then, and its memory lives on in both mine and the memories of my many friends who shared similar experience.

(Female, 1983)

Some were exposed to the kibbutz young generation's reflections and to those who are promoting the changes in the kibbutz way of life.

On the other hand, conflicting attitudes to the social ideals behind the kibbutz movement were also played out among the kibbutzniks, most markedly between the elderly kibbutzniks, the middle-aged and the young adults. Clearly, the younger generation (those in their 20s) wanted more individual liberty and a lesser degree of communal responsibility.

(Male, 1994)

There are some participants who are aware and updated to the current changes in the kibbutzim. They are even aware that there are different types of kibbutzim.

Of course, I have explained to the children that, unlike 'our' kibbutz, most of the kibbutzim are kibbutz in name only, that the concept, the ideal, is quickly vanishing with the harsh economic reality of today's world! Pity! yet, my friends and I are thankful everyday to have had the opportunity to share in some small way the reality that was kibbutz!

(Male, 1973)

Discussion and conclusion

Kibbutz volunteerism as multiple motivations – both altruistic and personal

While we did not attempt to ask our subjects to reconstruct their original motives for volunteering on kibbutz, it is clear that they represent a pluralistic mix of personal and ideological motives with differential and often unanticipated consequences on both dimensions. The respondents report both the search for novelty and escape to relaxation but also to a very high degree, the opportunity to build relationships both with hosts and volunteer peers. Not always did they find that their idealistic pre-trip image of the kibbutz was realized in the practice that they encountered, yet at the same time they were able to negotiate a new meaning to the experience as they lived through it, as well as in its recollection today.

Kibbutz volunteerism as an experience

The negotiation of asymmetry of power and meaning

While this self-selected sample of respondents is heavily weighted with satisfied volunteers, it is also clear that the asymmetry of the guest–host relationship did lead at times to the perception of exploitation and disappointment on behalf of some volunteers. Kibbutz volunteers could regard menial labor as fair instrumental exchange for the unlimited use of touristic recreation activities or could regard the same work as a let-down for the full social exchange anticipated by an ideologically motivated volunteer. By the same token, there is, and was, considerable variance between kibbutzim just as there is variance between types of volunteers, so that mismatches between the two could and did heighten the few antagonistic outcomes that were described here, both in the realm of work and in the type or absence of social relationships with hosts.

Characteristics of a good experience

Over and again, the reports above echoed all or part of the four dimensions that determine a good volunteer experience according to Pearce and Coghlan: 'Skill development and new personal insights, a good social life, experiencing novelty and contributing to a worthwhile project.' (2008: 136). Clearly for the vast majority of these respondents at least, the contribution to kibbutz life was a worthwhile experience in and of itself and it served the volunteers' own interests greatly. Kibbutz was new for most of them and the intensive social life with volunteer peers and frequently, but not always, with kibbutz-born youth peers, at a critical stage of emerging young adulthood, was a rich, pleasurable growing experience away from home.

The long-term impact of the experience in the development of the self

Indeed, it is quite remarkable how sustained the impact on self-development appears to have been on these kibbutz volunteer alumni. No matter which year they came, which gender and for the large majority of kibbutzim, these volunteers share a sense of social empowerment that many parents or schools would be proud to be able to engineer. What is the mechanism of self-development that is shared here over so many variables and over 30 years of experience? It seems to us that the secret lies in the universal attainment of interpersonal authenticity within the kibbutz volunteer experience, authenticity which was sought and found both with peers and with hosts.

Thus many years after the volunteer has left Israel and the kibbutz, has raised a family we find the dramatic finding (that is in line with previous research (Lev Ari and Mittelberg, 2008)) that kibbutz volunteer alumni repeatedly report a high degree of emotional attachment to the kibbutz society and to Israel so that similar to Noy (2004: 96), the experiences of tourists allow 'narratives of identity to be told, through the claim of a lasting self change'.

The data reported here reveal that the kibbutz volunteers were able to attain a significant degree of existential authenticity (Wang, 1999), within the framework of guest–host negotiation of the cross-cultural encounter in which they were engaged. Although they did not all come to the experience with the same background – far from it – nor did they interpret the experience in the same way, they were able to live through an existential state of authenticity that impacted their own identity. Their development of self was impacted by their experience with their volunteer peers and by their engagement with the kibbutz hosts, albeit in differing and perhaps unanticipated degrees.

Volunteers have their own needs and expectations as well as altruistic motivations and these vary over variables of demography and the type of experience offered by the host. As Pearce and Coghlan (2008) argue, interpersonal exchange theories contribute to our understanding of the volunteer experience; these in turn contribute to our understanding of the lasting impact of that experience on that self-development identified by Wearing. Thus a well-crafted volunteer experience can be a win–win situation: a period of critical growth and maturation for the participating volunteers as well as a boon for destination societies who can harness these growing energies towards the attainment of altruistic community centered goals, benefiting voluntourism itself, as well as the sending and receiving societies, at the one and the same time.

Acknowledgements

The authors would like to gratefully acknowledge the work of research assistant Efrat Cohen in the preparation of the qualitative material and our colleague Elliette Orchan in the original setting-up of the quantitative database.

References

Cohen, E. (1974) Who's a Tourist? *Sociological Review*, 22(4): 527–555.

Cohen, E. (1979) A Phenomenology of Tourist Experiences. *Sociology*, 13: 179–201.

Lev Ari, L. and Mittelberg, D. (2008). Between Authenticity and Ethnicity: Heritage Tourism and Re-ethnification Among Diaspora Jewish Youth. *The Journal of Heritage Tourism*, 3(2): 70–103.

MacCannel, D. (1992) *Empty Meeting Grounds: The Tourist Paper.* London: Routledge.

Mansfeld, Y. (1992) From Motivation to Actual Travel. *Annals of Tourism Research*, 19(3): 399–419.

Mittelberg, D. (1988) *Strangers in Paradise: The Israeli Kibbutz Experience.* New Brunswick (USA) and Oxford (UK): Transaction Publishers, pp. 223.

Mittelberg, D. (1999) *The Israel Connection and American Jews.* Westport Conn. USA: Praeger Publisher, pp. 216.

Noy, C. (2004) This Trip Really Changed Me: Backpackers' Narratives of Self-change. *Annals of Tourism Research*, 31: 78–102.

Palgi, M., Moyn, V. and Orchan, E. (1995) *The Kibbutz as a Path for Direct Absorption of Immigrants in Israel.* Institute for the Research of the Kibbutz and the Cooperative Idea, Haifa University (Hebrew).

Pearce, P.L. (2005) *Tourist behaviour: Themes and Conceptual Schemes.* Clevedon, UK: Channel View.

Pearce, P. L. and Coghlan, A. (2008) The Dynamics Behind Volunteer Tourism. In Lyons, K. D. and Wearing, S. (eds) *Journeys of Discovery in Volunteer Tourism.* Wallingford: CAB International, pp. 130–146.

Sutton, W. A. (1967) Travel and Understanding: Notes on the Social Structure of Touring. *International Journal of Comparative Sociology*, 8: 218–23.

Wang, N. (1999) Rethinking Authenticity in Tourism Experience. *Annals of Tourism Research*, 26: 349–370.

Wearing, S. (2001) *Volunteer Tourism – Experiences that Makes a Difference.* Oxford: CABI Publishing.

Part 2

Expanding the boundaries of volunteer tourism research

9   Volunteer tourism and divers with disabilities

Evidence from Malaysia

Caroline A. Walsh and Mark P. Hampton

Introduction

Volunteering and dive tourism is a rapidly expanding sector of tourism globally and is actively promoted as part of volunteering and volunteer tourism policy by the UK government and other Western governments (Rochester *et al*, 2009).

Dive tourism is rapidly growing in Malaysia. Malaysia regularly welcomes conventional leisure-based dive tourists to its world-class dive sites as well as those volunteering for scientific monitoring purposes, engaging in such activities as the Reef Check programme, as it has world-class diving locations. However, few disabled divers appear to participate in diving in Malaysia and, specifically, do not seem to be participating in volunteering and dive tourism. Consequently, this chapter discusses this based on findings from a series of in-depth semi-structured interviews. These semi-structured interviews were conducted online with internationally based disabled divers and also with dive instructors working in Malaysia. In addition, face-to-face interviews were also carried out in Malaysia. The interviews explored the issues of the wider context of disabled dive tourism together with the capacity of the dive industry in Malaysia to cater for disabled divers. Finally, the lack of participation of disabled divers in coastal and marine-based volunteering in dive tourism in Malaysia was examined.

Background

This research was undertaken alongside a major project measuring the developmental impacts of dive tourism on coastal communities in Malaysia. Until recently research into dive tourism has mainly focused on the physical impacts of diving on reefs and associated management issues (Musa, 2003). Virtually no research has been undertaken on the economic and social effects on local communities, especially in Malaysia. Globally, scuba diving is a fast growing activity; growing over 13 per cent year on year (PADI, n.d.) and global estimates indicate there are about 3–6 million divers.

The research that this chapter reports was undertaken at two dive sites, one in peninsula Malaysia, the other in Borneo, east Malaysia.[1] The Perhentian islands are small islands located off the east coast of the peninsula close to the

Thai border. Since the late 1980s the islands have attracted mainly international backpacker tourism, mostly for initial dive training (PADI Open Water certification) and also some 'fun dives'. Accommodation is mainly small scale and few large operators are yet located in the islands. The Perhentian islands have been subject to at least two formal volunteer tourism projects in the past, one of which was the Coral Cay Conservation Expedition in the early 2000s (Coral Cay Conservation, 2005).

Sipadan island is found off the east coast of Sabah, Borneo and mostly hosts international tourism through upmarket packages, although since around 2007 a new backpacker operator is now developing this market segment for scuba diving. Sipadan includes world-class dive sites and so attracts mostly experienced divers. There are no known formal volunteer tourism activities here.

Despite hosting several world-class dive destinations the numbers of disabled divers visiting and diving in Malaysia is unknown. Overall, the country is now a major tourism destination and hosts over 20 million international visitors each year (UN WTO, 2008). According to official government statistics two million tourists visit Malaysia to dive (Sabah Government, 2009).

Disabled dive tourism refers to dive tourism undertaken by disabled divers. Previous academic research has indicated that disabled individuals often have less social capital due to isolation (Aitchison, 2003). This is significant since it is recognised that social capital is needed to participate in tourism (Heimtun, 2007). Some individuals have had social capital but it has diminished due to acquired disability. In such cases resources from charitable institutions may allow them to participate within a constructed support group. This type of tourism is referred to as social tourism. Social tourism (Haulot, 1981; McCabe, 2009) seeks to overcome the barriers to tourism for excluded groups such as disabled groups by ignoring the underlying constraints and barriers, constructing an artificial economic environment of tourism through the application of the charity model, together with government regulation, interference and incentive for the tourism industry. This type of tourism has been recognised as not being sustainable particularly because the power relations are weighted against the disabled participants in favour of the organisers. The notion of free will is absent (Hall and Boyd, 2005).

There have been several initiatives worldwide to address the lack of empowerment affecting disabled people's everyday lives and life choices. Many of these initiatives aim to develop the specific types of capital held by the individuals, such as social capital. One initiative by the Handicapped Scuba Association uses diving to improve the social and economic position of individuals through engaging disabled individuals in a leisure education programme and dive tourism (HSA, n.d.). Elements have been replicated internationally, including in the UK by a disabled dive training organisation, the International Association of Handicapped Divers (IAHD) and the Back-Up Trust (Backuptrust, n.d.).

Another initiative in the UK has been the 'Volunteering for All' initiative. Until recently volunteering in the UK was something that was carried out 'on' disabled people, not *by* disabled people. There has been a campaign to change this perception in the UK, climaxing in 2005 with *The Time to Get Equal, Volunteering for*

All campaign by the charity SCOPE and its partners (SCOPE, 2005). Many of the barriers to volunteering have been institutional and societal. SCOPE's campaign focused on ways to attract disabled people to participate in all types of volunteering. It also raised awareness of the institutional and societal barriers that persist, whilst also proposing some solutions. Stebbins (2007) attempted to promote lifestyle volunteering for those disabled people who cannot work and therefore, who are at risk of social isolation – which can then lead to less social capital and fewer opportunities in a negative reinforcing loop. Stebbins (2007) has been equally vocal about promoting serious leisure and associated activities to the same disabled groups and recommending methods to encourage this. He has largely focused on those participating in rehabilitation, or who have learning disabilities.

The issue of volunteering in tourism by disabled people has not been explored in the academic literature. The highly cited definition of volunteer tourism by Wearing (2001: 1) allows for a spectrum of volunteer activities to be considered under the volunteer tourism banner. This ranges from the informal volunteer activity taken as part of a leisure holiday where no money is paid for the volunteering experience, to the other end of the scale of organised volunteer tourism activities undertaken as part of serious leisure, whereby the individual(s) pay an organisation for the experience. It is argued that this definition is too narrow and does not include the element of financial payment by the volunteer. The payment by volunteers to volunteer organisations to attend organised volunteer activities is only one end of the volunteer tourism scale. It is, however, accepted that these organised volunteer activities are those which have been focused upon exclusively in the academic literature. Stebbins (1992) classified volunteer tourism as part of 'serious leisure'. He explored the issue of under-representation in serious leisure and how to address this and the benefits to disabled individuals who take part in serious leisure. He does not address or discuss participation in serious leisure by those persons with disability who have not undertaken some form of rehabilitation.

To paraphrase Wearing (2001), volunteer tourism is a holiday/type of tourism with a specific focus and purpose for good/change/making a difference. Therefore, dive volunteer tourism is dive tourism with a specific purpose or focus that is making a contribution to the conservation of marine or coastal environs. Coghlan (2005) has identified the range of dive volunteer tourism projects and their variety of focus. Unfortunately, there is still a lack of evidence concerning the involvement of disabled divers in this type of volunteering.

Methodology

The data collection took place in April and May 2009 and triangulation was conducted between dive professionals and disabled divers. The possible population of respondents was initially located using one of the authors' existing contacts from the *Facebook* social networking site. These were then used to 'snowball' to lead to further contacts that had been recommended by the initial interview respondents (Brickman-Bhutta, 2009). The online in-depth semi-structured interviews were carried out with internationally based disabled divers and dive instructors working in Malaysia. In Sipadan, face-to-face interviews with dive instructors and dive operators were also completed. The online interviews with disabled divers were conducted in two sessions. There were n=13 respondents to the first questionnaire and n=12 to the second. There were n=7 respondents from the dive professionals. Of the thirteen disabled divers that were surveyed, nine were from the UK, two from the USA and two from Hong Kong. All were surveyed through snowballing sampling. Eleven had acquired physical disability within the last fifteen years and none had learned to dive prior to that. Only two respondents surveyed had been disabled from childhood, and of these, one had been disabled from birth. All of the UK disabled divers had learned to dive as part of a leisure programme during ongoing physical rehabilitation. Both of the divers from USA had had diving recommended to them as a suitable leisure pursuit after acquiring their disability.

Results and discussion

Of the thirteen disabled divers approached, only one, one of the US disabled divers, had considered formal volunteer tourism but had not been encouraged to participate on the grounds of her physical disability by an environmental organisation. Of the remaining respondents there were four main categories of reply:

1   participating in informal volunteer activities as part of their dive tourism experiences, such as informal reef clean up, and did not realise these types of activities could be considered volunteer activities
2   training to volunteer within the dive industry context
3   are aware of volunteer activities but do not know whether they can participate
4   are unaware of what volunteer tourism or volunteering is.

All respondents mentioned that seeing and experiencing the underwater environment was a motivation to continue diving. None were explicitly hostile to the possibility of participating in volunteer tourism in the future. Two of the disabled divers from Hong Kong had visited Malaysia. Of the remaining divers, only one suggested that Malaysia was 'on their list of places to visit' citing a lack of finances for not visiting before. The lack of specifically mentioning Malaysia as a key dive destination as a 'must' visit by other respondents is surprising as Sipadan is billed as one of the ten 'must see' dive destinations in the world within the diving community. A few of the respondents did state that there were many places that they would like to visit and they could not remember them all, so the fact that Malaysia and specifically Sipadan was not mentioned may not be as significant as it first seems. Follow-up interviews would need to be undertaken to clarify this point.

All dive professionals interviewed or surveyed from Malaysia felt that where they were working was not physically equipped to cater for disabled divers generically and, therefore, even less able to cater for their needs during volunteer tourism projects. However, they did emphasise that they would not stop disabled

divers from participating in such activities if they were offered at their dive establishments and they believed at least some of the dive professionals on the staff
in their dive establishment had a positive attitude to overcoming those structural
barriers to participation. Only one dive professional placed the onus back on the
disabled diver with remarks 'if they can with minimal help' (Respondent DO2).

All of the Malaysian instructors interviewed believed that it was culturally
their duty to serve disabled people and assist them as much as they can. One of
the Malaysian dive operators interviewed choose to quote the PADI dive training
manual chapter verbatim rather than express his own views (Respondent DO2),
which were contrary to the views of the other Malaysian instructors interviewed
(Respondents DI1–DI4). Interestingly, the international instructors surveyed
believed that the international dive professionals would be better equipped skill
wise and in attitude to cater for disabled divers due to their exposure to disability discrimination laws and awareness in their home countries but the Malaysian
instructors would be restricted by their culture (Respondent DIP2). This is contrary to the personal experience of one of the authors whilst diving at least some
of these locations.

After analysing the responses of the interviews it is clear that the constraints
identified can be classified into three broad categories: (1) physical or structural
barriers; (2) self-belief (or lack of), motivation, or perception to participation and
(3) perceived or actual inter relations between disabled divers and dive tourism
providers or other non-disabled divers.

These identified categories replicate those identified in the leisure constraint
theory. Leisure constraint theory has developed to explore issues relating to participation of leisure activities including tourism. The leisure constraint models
are the backbone to the leisure constraint theory and have been used to explain
constraints of specific groups to participate in specific types of tourism. Examples
include issues relating to older tourists (Lee and Tideswell, 2005), ski tourists
(Hudson and Gilbert, 1999) and nature tourists (Pennington-Gray and Kerstetter,
2002; Nyaupane *et al.*, 2004). There appears to be no studies related to leisure
constraint theory and disabled divers; the closest literature is by Daniels *et al.*
(2005) who examined disabled tourism and Todd *et al.* (2001) who researched the
leisure constraint theory in respect to Scuba divers generically. Leisure constraint
models have been developed considerably since Crawford *et al.* proposed their
first model in 1991. This original model had three main components, classing
constraints as *interpersonal, intrapersonal* and *structural. Interpersonal* denotes
anything that interacts with the individual; *intrapersonal* being constraints caused
within the individual themselves such as their attitude. *Structural* refers to the
physical environment or something of a physical nature. These categories were
purposefully left somewhat wide to cater for a wide-range of excluded groups
and their related issues; however, this has been widely criticised by Samdahl
and Jekubovich (1997). The model has been added to and operationalised and
now includes such fields as motivation and negotiation. The three fundamental
building blocks of the original model remain. In addition, the model explicitly
has other elements that are not categorised under the categories of interpersonal,

intrapersonal and structure categories. These have been identified as acting on or influencing the three main components, influencing the outcome of whether participation takes place and the type of participation. These additional components include: motivation; negotiation strategies and facilitators. Motivation literature is extensive and, consequently, needs little discussion here. An example of a negotiation strategy is the ability of disabled divers to discuss with dive tourism providers and mitigate any barriers identified. The facilitators' category is a very recent addition, only being operationalised since 2007. This development is in response to criticism that in some cases the constraints are not the overriding issue that influences the participation but the presence or absence of facilitators are. A facilitator can be defined as something that acts as a catalyst to allow something else to happen. Raymore (2002) observed that because constraints are not present does not automatically lead to participation. Facilitators include such examples as: 'does the individual attend a leisure education programme as part of rehabilitation'. In addition to adding the facilitator aspect to the model, Raymore (2002) also proposed that the model should no longer be hierarchical but take on a spherical structure. Raymore (2002) argues that the hierarchical nature of the model had been criticised for some time. Indeed, Daniels *et al.* (2005) had already highlighted that in respect to disabled people and measuring constraints to activities, specifically travel, the model should be interactive and not hierarchical.

Also identified in the literature are 'contested influencing factors', namely: coping strategies and personality. Coping strategies are viewed as contentious as seen by the debate sparked by Schneider *et al.* (2007). In response to their original paper the debate that ensured, in part, argued that incorporating the element of coping strategies risks harbouring back to the medical model of disability and other models of social exclusion and not in keeping with the more favoured social model. Coping strategies have their merits and explain some kinds of behaviour whereby negotiation to overcome barriers does not take place. For example, people may choose to employ a coping strategy to hit the constraints head on (Son *et al.*, 2008). McGuiggan (2004) added to leisure constraint theory by proposing a vacation choice model that combined the leisure constraint model and personality. She argued that personality bears an influence on holiday preference by affecting motives and the perspectives of the constraints that may actually or may be perceived to exist as well as an individuals ability to overcome these.

Whichever structure the model takes, it is also acknowledged that one or more influencing factors may work together to influence positively or negatively on the constraints to, reduce or reinforce them (Raymore, 2002). It is also recognised in the literature that specific influencing factors may work against each other to also influence negatively or positively on constraints. This aspect has now been operationalised and there is wide application in the literature based on Mannell *et al.*'s (2005) work. Carroll and Alexandris (1997) argue the need for greater understanding of relationships between perceived constraints, motives and motivation. Alexendris *et al.* (2002) argue that their study demonstrated how the motivation dimensions impact, interact and influence on interpersonal constraints and thus behavioural actions resulting in either non-participation or participation as also demonstrated earlier by Crawford *et al.* (1991). Alexendris *et al.* (2002) also argued that the self-determined model (Deci and Ryan, 1985) and multi-dimensional of motivation model (Vallerand and Losier, 1999) are useful theoretical frameworks to use in order to study relationships between constraints and motivations.

The academic literature broadly suggests that as divers become more qualified and more experienced they become more environmentally aware. The environmental and coral reef conservation academic literature discusses this in more detail within the context of environmental behaviour. The literature (Dearden *et al.*, 2006; Thapa *et al.*, 2005) suggests that, in high-income countries at least, there is a type of 'career ladder' relating to the development of environmental behaviour amongst divers[2] as follows:

1   Environmental awareness i.e. coral reefs are not rocks but living organisms and need protection.
2   Environmental concern e.g. coral reefs are under threat from human activity and action is required.
3   Environmental action e.g. taking part in clean up dives, Reef Check to monitor the impact on the reef by human activity, gravitating towards volunteer tourism projects that are more organised and do environmental research as a form of 'hard core' action.
4   In some cases this may lead to environmental professionalism.

There is no evidence that disabled divers follow this same trend. Assuming that the environmental 'career ladder' is reasonably universal, this then raises a question to be addressed, that is, why is it that disabled dive tourists do not get to the stage in their environmental 'careers' where they take action (volunteer tourism)? Is it because they do not gain enough dive experience, have sufficient qualifications, or enough appropriate experience of environmental education to get to this stage? Is the reason why disabled divers do not seem to follow the above-mentioned trend because they come to diving from a different position? That is, diving was initially a form of rehabilitation as opposed to recreation. If so this implies that they might not then be trained to higher levels. This present research reported in this chapter would not support this conclusion. It was clear that all but one of our respondents (who was a dive trainee) had achieved at least Rescue Diver, or higher level, diving qualifications. All the UK and USA respondents had learned to dive as part of a rehabilitation programme, acting as a facilitator to participation. In the UK there was some differences and respondents regularly took part in diving holidays organised by themselves as well as fellow disabled divers; in these cases, this moves beyond the framework of rehabilitation and into diving for recreation purposes. The dive destinations visited included the Red Sea, the Caribbean and Asia. This seems to indicate a diverse range of experiences and individual negotiation strategies are at play when these divers embark on a diving holiday. Also, their motivation is beyond that of just rehabilitation. This is supported by the responses in the research. All

UK divers indicated that they partake in diving for the sense of freedom and to experience the underwater environment. The US respondents' experiences seemed very different and on the whole were arranged through others or disability organisations *for them* and they explicitly stated their dive experiences were limited by the decisions and perceptions of others. Our research would suggest that there are rather more complex and cultural aspects at play than earlier theories suggest; this needs to be explored further before a firm conclusion can be made regarding this question. Cultural and personality aspects were not investigated as part of this exploratory study.

In the US, disabled dive tourism has developed as a specialist niche of social tourism. It is unclear that if volunteer tourism was developed within the social tourism framework or as an extension of it, and therefore, would disabled divers participate? It is not clear from our research whether this is the case; however, what is apparent from the responses is that organised dive tourism does take place amongst disabled divers who participate in organised leisure education programmes as part of their physical rehabilitation. In the UK this type of tourism does not fulfil the criteria of social tourism as it is organised by disabled divers for disabled divers on an informal basis and does not construct an artificial environment of any kind. Some of the trips organised by some of the disabled dive tourism organisations in the US follow a social tourism framework. Promoting volunteering as an activity that takes part of their dive tourism was posed as an interview question and received unanimous approval by respondents. Nevertheless, the lack of awareness and organisation of such activities exclusively for disabled divers, or integrating disabled divers was identified as an obstacle to participation.

Coghlan (2005) argues that there is a notable trend amongst volunteer tourists in the field of marine tourism of what can be called 'career volunteerism' where volunteers often move from volunteer to volunteer expedition staff roles. Some later move on to pursue environmental careers. This raises the question: why is career volunteering not occurring amongst disabled divers? Stebbins (2004) suggests that career volunteerism does occur amongst some disabled people. All but one respondent in this research expressed a lack of awareness of volunteer tourism and related opportunities and none mentioned a negative perception of volunteering as a reason for a lack of participation.

There is evidence in the literature that a diver's decision to participate in volunteer tourism is influenced by their own personal dive education (Dearden, 2006). Dive education is influenced by the dive professionals that divers are trained by. This includes their instructor's attitudes, perspectives and capacity to cater for them and the facilities on offer generally which, in turn, influences their educational experience. In the case of disabled divers this is magnified in respect to the structural and attitudinal (perceived or actual) barriers that they also encounter. Both divers and instructors in our research also identified the physical environment at dive locations as barriers.

Todd *et al.* (2001) argues that motivation to do something is influenced by the level of past experiences. This raises the question of whether the past experiences of disabled divers adversely affect their motivation to get involved in volunteer tourism. This is not clear from this research as only one respondent explicitly stated that they had enquired about partaking in volunteer tourism with a negative reaction. All but one diver (who was currently training) seem to have wide dive experiences, visiting a range of destinations but many do choose to return to one or two dive locations on a regular basis. Therefore, the location may be an issue. Further research is required to establish this. Todd *et al.* (2001) also argues that motivation can be explained by the expectancy-value theory which states that motivation is determined by the attractiveness of the outcomes and expectation that participation will result in desired outcomes. So why, all things being equal, would the outcomes of participation in volunteer tourism be less attractive to disabled divers? Can leisure constraint theory explain this? Despite the diving experiences of the divers surveyed, the question remains: do disabled divers modify their participation in diving consciously or subconsciously compared to their able bodied counterparts and therefore have differing expectations of participation and the outcomes from participation in diving related activities? It is unclear from this research if this is the case. Todd *et al.* (2001) also argues that the more qualified the scuba diver the less that constraints are experienced. It is unclear whether the lack of participation by disabled divers was due to the lack of appropriate qualifications. The results of this study would not support this. As already discussed the majority of the disabled divers interviewed had achieved Rescue Diver certification,[3] and were actively engaging in further training.

Conclusions

This chapter has explored the influencing factors on participation in volunteering in dive tourism, and the experience of disabled divers who seemingly have the same motivation and experiences as their able-bodied peers.

Both disabled divers and instructors alike identified major constraints as barriers to participation. However, both groups offered solutions to these limitations. Therefore, other types of factors appear to be interacting with the structural constraints to restrict participation. The types of constraints identified by both groups in our research are similar to the constraints affecting participation in other types of tourism. A larger study on a more international scale is needed of both instructors and disabled divers to clarify our initial findings. Operationalising the leisure constraint theory within the context of volunteering in dive tourism would be beneficial to explore influencing factors in more depth.

Acknowledgements

The authors are grateful to the British Council for its PMi2 Research Cooperation Award 2008–10 for the research on dive tourism impacts in Malaysia. The authors would also like to thank all the interview respondents who generously gave up their time. The authors would like to acknowledge and thank Dr Ghazli Musa for his input into the research design of the research in this chapter. The usual disclaimers apply.

Notes

1 Tioman Island was also considered. Tioman Island was also considered as a possible research site but scuba activity there is broadly a combination of the two main types seen in the other two islands: initial dive training (as in the Perhentian Islands) and scuba diving by more experienced divers, often staying in more upmarket accommodation (Sipadan). Volunteer tourism also appears to be in its infancy in Tioman.
2 It is unclear whether this same 'career ladder' is also found in less developed countries given the perhaps more limited opportunities and lower disposable income of many local people to progress through increased training in scuba diving.
3 Although the different dive training organizations (PADI, BSAC, NAUI etc.) use slightly differing terms for their training levels, broadly dive certification is graded into common levels from initial certification (Open Water Diver), via intermediate levels (Advanced Open Water Diver, Rescue Diver) through to most highly trained (Dive Master, Instructor).

References

Aitchison, C. (2003) From leisure and disability to disability leisure: developing data, definitions and discourses. *Disability and Society*, 18(7): 955–969.

Alexandris K., Tsorbatzoudis C., and Grouios G. (2002) Perceived constraints on recreational sport participation: investigating their relationship with intrinsic motivation, extrinsic motivation and amotivation. *Journal of Leisure Research*, 34: 233–252.

Back-Up Trust (n.d.) *SCUBA Dive with BackUp Trust*. Available at www.backuptrust.org. uk/scuba.html. Accessed 17 December 2009.

Brickman-Bhutta, C. (2009) *Not by the Book, Facebook as a Sampling Framework*. Available at http://www.thearda.com/workingpapers/download/Not%20by%20the%20 Book%20-%20Bhutta.doc. Accessed 19 December 2009.

Carroll, B. and Alexandris, K. (1997) Perception of constraints and strength of motivation: their relation to recreational sport participation. *Journal of Leisure Research*, 29: 279–299.

Coghlan, A. (2005) *Towards an Understanding of the Volunteer Tourism Experience*. Unpublished thesis, James Cook University, Townsville.

Coral Cay Conservation (2005) *Malaysia Expedition Reports*. Available from http://www. coralcay.org/component/option,com_docman/task,cat_view/gid,124/Itemid,393/. Accessed 17 December 2009.

Crawford, D., Jackson, E. and Godbey, G. (1991) A hierarchical model of leisure constraints. *Leisure Sciences, 13: 309–320.*

Daniels, M. J., Drogin Rodgers, E. B. and Wiggins, B. P. (2005) Travel tales: an interpretive analysis of constraints and negotiations to pleasure travel as experienced by persons with physical disabilities. *Tourism Management*, 26(6): 919–930.

Dearden P., Bennett, M. and Rollins, R. (2006) Implications for coral reef conservation of diver specialization. *Environmental Conservation*, 33: 353–363.

Deci, E. L. and Ryan, R. M. (1985) *Intrinsic Motivation and Self-determination in Human Behavior*. New York: Plenum.

Hall, C. M. and Boyd, S. (2005) (eds) *Tourism and Nature-based Tourism in Peripheral Areas: Development or Disaster*. Clevedon: Channelview Publications.

HSA (Handicapped Scuba Association) (n.d.) http://www.hsascuba.com. Accessed 18 December 2009.

Haulot, A. (1981) Social tourism: current dimensions and future developments. *International Journal of Tourism Management*, 2(3): 207–212

Heimtun, B. (2007) From leisure and disability to disability leisure developing data, definitions and discourse. *Disability and Society*, 18(7): 955–969

Hudson, S. and Gilbert, D. (1999) Tourism demand constraints: a skiing participation. *Annals of Tourism Research*, 27: 906–925.

Lee, S. and Tideswell, C. (2005) Understanding attitudes towards leisure travel and constraints faced by senior Koreans. *Journal of Vacation Marketing*, 11(3): 249–263.

Mannell, R. and Iwasaki, Y. (2005) Advancing quantitative research on social cognitive theories of the constraint-negotiation process. In E. Jackson (ed.) *Constraints to Leisure*. Pennsylvania: Venture Publishing, pp. 261–278.

McCabe, S. (2009) Who needs a holiday: evaluating social tourism. *Annals of Tourism Research*, 36(4): 667–688.

McGuiggan, R. L. (2004) A vacation choice model incorporating personality and leisure constraints theory. *Tourism Analysis*, 8(2–4): 187–191.

Musa, G. (2003) Sipadan: an over-exploited scuba-diving paradise? An analysis of tourism impact, diver satisfaction and management priorities. In B. Garrod and J. C. Wilson (eds.) *Marine Ecotourism: Issues and Experiences.* Clevedon, UK: Channel View, pp. 122–137.

Nyaupane, G., Morais, D. and Graefe, A. (2004) Nature tourism constraints: a cross-activity comparison. *Annals of Tourism Research*, 31(3): 540–555.

PADI (Professional Association of Diving Instructors) (n.d.) http://www.padi.com. Accessed 18 December 2009.

Pennington-Gray, L. and Kerstetter, D. (2002) Testing a constraints model within the context of nature based tourism. *Journal of Travel Research*, 40: 416–423.

Raymore, L. A. (2002) Facilitators to leisure. *Journal of Leisure Research*, 34(1): 37–51.

Rochester, C., Paine, A. E., Howlett, S. with Zimmeck M. (2009) *Volunteering and Society in the 21st Century*. Hampshire: Palgrave Macmillan.

Sabah Government (2009) http://kepkas.sabah.gov.my/index.php?option=com_contentan dview=articleandid=9822:expo-to-encourage-diving-among-locals-andcatid=42:year-2009andItemid=132. Accessed 17 December 2009.

Samdahl, D. and Jekubovich, N. (1997) A critique of leisure constraints: Comparative analysis and understandings. *Journal of Leisure Research*, 29(4): 430–452.

Schneider, I. E. and Stanis, S. A. W. (2007) Coping: an alternative conceptualization for constraint negotiation and accommodation. *Leisure Sciences*, 29(4): 391–401.

SCOPE (2005) *Time to get equal. Banishing disablism in volunteering*, SCOPE, London.

Son, J. S., Mowen, A. J. and Kerstetter, D. L. (2008) Testing alternative leisure constraint negotiation models: an extension of Hubbard and Mannell's model. *Leisure Sciences*, 30: 1–19.

Stebbins, R. A. (2004) *Volunteering as Leisure/Leisure as Volunteering*: An International Perspective. Wallingford: CAB International.

Stebbins, R. A. (2007) *Serious Leisure: A Perspective for our Time*. Available from Aldine New Brunswick, NJ: Transaction Publishers.

Thapa, B., Graefe, A. R. and Meyer, L. A. (2005) Moderator and mediator effects of scuba diving specialization on marine-based environmental knowledge-behavior contingency. *Journal of Environmental Education*, 37(1): 53–67.

Todd, S., Graefe, A. and Mann, W. (2001) Differences in scuba diver motivations based on level of development. In S. Todd (ed.) *Proceedings of the 13th Northeastern Recreation Research Symposium.* Gen Tech Rep. NE-289. Newton Square, PA: US Department of Agriculture, Forest Service, Northeastern Forest Experiment Station. pp. 107–114.

UN WTO (2008) *Yearbook of Tourism Statistics Data 2002–2006*, http://pub.unwto. org/WebRoot/Store/Shops/Infoshop/4868/9DE3/BC9B/D054/FF6C/C0A8/0164/ 1115/080630__0020_yearbook_2008_excerpt.pdf. Accessed 19 December 2009.

Vallerand, R. and Losier, G. (1999) An integrative analysis of intrinsic and extrinsic motivation in sport. *Journal of Applied Sport Psychology*, 11: 142–169.

Wearing, S. (2001) *Volunteer Tourism: Experiences That Make a Difference.* Wallingford: CABI Publishing.

10 Profiling volunteer holiday leaders

A case study of National Trust working holiday leaders – socio-demographics, basic human values and functional volunteer motivations

Steven Jackson

Introduction

The National Trust (NT), a UK-based charity with a remit for the preservation and conservation of natural and built heritage, has run working holidays for over forty years. These holidays consist of volunteer participants, volunteer leaders and local NT staff. The leaders play an important role as the link between the participants, local NT staff and the wider NT organisation. They also perform a broad range of tasks requiring a variety of skills, abilities and knowledge (see Table 10.1). Most leaders have a long-term commitment to the role spanning several years with participation in a number of holidays each year. There is a lack of research in volunteer tourism concerning this long-term allegiance to an organisation and limited work on leaders of volunteer holidays (e.g. Coghlan, 2008), let alone leaders who may be volunteers themselves. Additionally, although values have been paid lip service in relation to volunteer tourism (e.g. Halpenny and Caissie, 2003) or have formed the basis of hypothetical models (e.g. Lockstone *et al.,* 2002), there has been a lack of thorough analysis. Values have, however, been extensively examined in relation to the environment (e.g. Stern and Dietz, 1994; Schultz and Zelezny, 1999; Aoyogi-Usui *et al.,* 2003). Similarly, motivations to volunteer have concentrated on immediate, proximate reasons for volunteering without consideration of more fundamental motivational factors or immediate antecedents or distal influences. This chapter, therefore, sets out to answer a number of questions:

1 Who is it that takes on the role of volunteer leader?
2 What are the basic human values of these volunteer leaders?
3 What are the fundamental motivational factors that lead them to volunteer?
4 How are the above three areas related to each other?

Steven Jackson

Table 10.1 The role of the National Trust Working Holiday Leader

Purpose of role:	To host a National Trust residential working holiday, facilitate team cohesion in a group of around twelve volunteers whilst maintaining a safe work and domestic environment on a National Trust working holiday.
Key elements:	1 Welcoming holiday participants on behalf of the property and the National Trust. Create an environment which promotes the emotional, physical and spiritual well-being of all the participants on the holiday and which encourages participants to play a full and active part in the holiday.
	2 Planning and organising domestic arrangements, including arrivals and the catering budget, ensuring hygiene and health and safety instruction are followed in the accommodation. Facilitating the organisation of the menus, cooking, social programme and general chores in the accommodation.
	3 Managing the group to ensure that everyone is equipped and on the work site at the appointed time, assisting Trust staff in briefing, working with, supervising and motivating the group on site and ensuring that safe working practices are adopted in accordance with the size and abilities of the group.
	4 Develop and promote team building and social interaction within the group and between group members and National Trust staff.
	5 Making appropriate provision for all holiday participants to achieve their holiday goals, including Duke of Edinburgh and John Muir awards.
	6 To act as an ambassador for the National Trust to holiday participants and the general public. Actively promote current Trust policy, e.g. initiatives regarding climate change.
	7 Nurturing and developing leadership potential in new leaders and with volunteers who aspire to become leaders in the future.
	8 Being aware of any medical requirements and special needs of the group while being prepared to deal with any injury or illness promptly and effectively.
	9 Taking part in regular Working Holiday leader events in order to keep up to date with any changes in Working Holidays procedure and wider Trust developments.
	10 Undertaking such other appropriate assistance in relation to hosting working holidays as may be reasonably requested by the National Trust manager.
	11 Working in compliance of the Trust's Health and Safety Policy at all times.

Source: *Working Holidays Leaders Newsletter*, Spring 2008, The National Trust, Swindon

The National Trust was established in 1895 in response to a concern about uncontrolled industrial development and its impact in the UK (full details of the National Trust, including the following information, are available from their website: www.nationaltrust.org.uk). The National Trust is a registered charity independent of government and is considered one of the major conservation organisations in the UK. It has 3.5 million members and 43,000 volunteers. It cares for a vast estate consisting of over 200 houses and gardens, 40 castles, 12 lighthouses and 6 World Heritage Sites. In addition, it protects over 700 miles of coastline and over 250,000 hectares of countryside. In 2007, pay-for-entry properties attracted 12 million people and an estimated 50 million people visited the Trust's countryside properties.

As indicated above, the National Trust has an extensive volunteer programme. This includes full-time volunteering, employee volunteering, volunteer groups, youth volunteering and working holidays. The working holiday programme has been in existence for over 40 years. Working holidays usually consist of about a dozen volunteers. The cost of the holiday includes all accommodation, transport after arrival, food, tools and protective equipment. Accommodation is usually in specially converted base camps (although other types of accommodation may be used) and volunteers sleep in dormitories. Volunteers cater for themselves under the guidance of the leaders (who devise the menus and purchase the food). Volunteers arrive by their own transport or are met by the leaders at a local public transport point. Saturday evening is spent in introductions by the leader (and sometimes the local National Trust staff) and settling in. Sunday is often an orientation day with an introduction to the work site, the work to be undertaken, any issues regarding health and safety, and any training that may be required. Subsequent days are spent on the work tasks (with Wednesday as a free day) and departure on the following Saturday. Until the beginning of 2008, working holidays had been classified into a variety of categories based on age and type of work. The majority of the work is concerned with physical conservation tasks such as scrub clearance, removal of invasive species, footpath maintenance and drystone wall building. In addition, some working holidays are based in gardens while some holidays are more specialised and involve biological survey work, archaeological excavations, assisting at open air events or working with volunteers with disabilities.

Working holiday leaders have an important role to play in the process of facilitating working holidays (references to leader includes assistant leader). The position is voluntary, but the normal holiday cost is waived and transport costs to and from the holiday are paid. In addition, there are reasonable allowances for telephone and postage prior to the holiday. Holidays of a week's duration have a leader and assistant, those over a weekend a leader only. The leaders are the link between the volunteers taking part on the holiday and the National Trust staff who are responsible for the work carried out on the work site. The role of leader (see Table 10.1) covers an extremely wide range of functions from contacting volunteers prior to the holiday and basic care for the holiday volunteers, through to motivation, budgeting, social functions, being an ambassador and driving the minibus or Land Rover. Clearly, the role of leader is often pivotal to the success of the working holiday and depends on a wide range of management and leadership skills.

Given the above information regarding the role of National Trust working holiday leaders, the question arises as to who volunteers for this role and why? Such information is valuable for volunteer managers in recruiting and retaining volunteer leaders and 'understanding the motives of different volunteers will provide volunteer managers the opportunity to effectively promote opportunities and design volunteer positions that fulfil the interests of potential volunteers' (Yoshioka *et al.*, 2007: 34). Additionally,

> continued participation depends on the person-situation fit, such that volunteers who serve in roles that match their own motivations will derive more satisfaction and more enjoyment from their service and be more likely to intend to continue to serve than those whose motivations are not being addressed by their activities.
>
> (Clary *et al.*, 1998: 1528)

Moreover, Wymer (1997: 22) suggests that studies of values 'may give managers ideas about what messages to emphasize in volunteer recruitment appeals as well as what messages not to emphasize' and that 'understanding what values are important and unimportant to an organization's volunteers may give managers insights about how to provide volunteers with satisfying experiences, possibly increasing volunteer retention'.

Building the argument

Within this section the characteristics of volunteer tourists and environmental volunteers are briefly outlined before examining the areas of values and motivations. Finally, a simple model is presented to link the above three areas.

Weiler and Richins (1995: 34), studying participants of Earthwatch expeditions, collected data on gender, marital status, age, education and income, concluding that 'the typical Earthwatch Team Member is female, single, between the ages of 26–35, well-educated, well-paid and professional'. This contrasts markedly in terms of age with the Habitat for Humanity (Stobbart and Rogerson, 2004) sample who were predominantly in their 20s or 50s, but shows a similar concentration in the professional and managerial occupational categories with the addition of a large student category, presumably those in their 20s. Operation Wallacea appears to attract this younger, student category, Galley and Clifton (2004: 74) finding that the volunteers were 'predominantly female (63 per cent), British, single and aged between 20 and 22' with the largest number being students studying for a science or environmentally based degree at university. This younger age category was also represented in Broad's (2003) study of a gibbon rehabilitation project where the average age was 25, again predominantly female, and predominantly from Europe. The professional orientation of volunteers is a recurring theme, as illustrated by Campbell and Smith (2006) in a study of volunteer tourists involved in turtle conservation in Costa Rica. Most volunteers were in professional occupations or at university training for the professions.

These findings for volunteer tourists may be compared with similar data for 'resident' volunteers in environmental programmes. Bradford and Israel (2004), for example, again in relation to turtle conservation (in Florida), found that two-thirds of the sample was female, the average age was 53, and they were well educated and white. Forty-five percent of the sample did not work for pay. A study of an environmental stewardship group in Canada by Donald (1997) found that the largest age group was the 36–49 year-olds, but again from a well-educated, white, European background. Some demographic characteristics appear to transcend cultures. Wu (2002), in an examination of volunteer interpreters for Taiwan's national park, found that the main age group was 26–40 and well educated with a college or graduate school education. Using data from the World Values Survey for 1999–2002, Randle and Dolnicar (2006) show that for environment and animal rights volunteers, the majority were male, married, with children, in full-time work and middle-class.

Values are often seen as the foundations on which human society is organised; they form the base of values-attitudes-behaviour models (e.g. Vaske and Donnelly, 1999; Manfredo and Dayer, 2004; Jackson, 2007) and have been recognised as the core of personal identity (Hitlin, 2003). Values have recently been reviewed as a general concept (Hitlin and Piliavin, 2004) and specifically in relation to the environment (Dietz, Fitzgerald and Shwom, 2005). Although much of the early empirical work stems from the studies by Rokeach (1973), concentration here is on the work of Schwartz (1994) and co-workers whose value system has been used in a wide variety of contexts. Schwartz (1994: 21) defines values as

> desirable transsituational goals, varying in importance, that serve as guiding principles in the life of a person or other social entity. Implicit in this definition of values as goals is that (1) they serve the interests of some entity, (2) they can motivate action – giving it direction and emotional intensity, (3) they function as standards for judging and justifying action, and (4) they are acquired both through socialization to dominant group values and through the unique learning experiences of individuals.

The present study is concerned with the motivational role of values and as such the 'primary content of a value is the type of goal or motivational concern it expresses' (Hitlin, 2003: 119). Thus, 'values are probably less important as predictors of whether somebody volunteers, but more of an aid to understanding what kind of reasons and motivations are appealing' (Dekker and Halman, 2003: 7).

However, Hitlin (2003: 125) points out that 'The values at the core of the self that produce our sense of personal identity are distal influences on action, shaping and channelling the choice of situations in which we interact'. Thus, Bardi and Schwartz (2003: 1207) argue 'that a single motivational structure organizes the relations among the sets of values and behavior' but also suggest that 'It is still unclear whether values relate to behavior generally or only that some values relate to some behaviors.' It needs to be stressed, however, that (Dekker and Halman, 2003: 5) 'General motives are also very different from tangible reasons for participating' [in volunteering].

Such tangible reasons have often been the core of studies of motivation for taking part in volunteer tourism (Jackson, 2009). These range from those that may be considered more altruistic such as giving back to nature or doing something meaningful in terms of conservation (Broad, 2003; Brown and Lehto, 2005) or helping those less fortunate (Stoddart and Rogerson, 2004), to those that are clearly aimed at self-development perhaps in terms of career (Galley and Clifton, 2004), and those that are more social such as meeting like-minded people (Stoddart and Rogerson, 2004). In some instances there is the motivation of adventure (Stoddart and Rogerson, 2004) or challenge (Galley and Clifton, 2004) or the wish to be immersed in the local culture (Brown and Lehto, 2005; Clifton and Benson, 2006). These studies, however, are often concerned with motivations directly connected to the individual projects, without an examination of deeper, more fundamental motives. They also adopt their own individual approaches with a variety of methods, ranging from focus groups to interviews and questionnaires. As a consequence it is difficult to make more than generalisations concerning the similarity of motivational factors. However, the functional approach to understanding volunteering is becoming increasingly popular as it applies to the motivations for undertaking volunteer work (Clary *et al.*, 1998).

Table 10.2 Schwartz's second order values and summary statistics for volunteer leaders in the present study

Type	Statement	Mean	St. dev.	N	Missing
Universalism	Understanding, appreciation, tolerance, and protection for the welfare of all people and nature	5.0	1.05	183	39
Self-direction	Independent thought and action – choosing, creating, exploring	4.9	1.06	183	39
Security	Safety, harmony and stability of society, relationships and oneself	4.7	1.14	183	39
Stimulation	Excitement, novelty, and challenge in life	4.6	1.14	182	40
Benevolence	Preservation and enhancement of the welfare of people with whom you are in frequent personal contact	4.4	1.15	184	38
Tradition	Respect, commitment, and acceptance of the customs and ideas that traditional culture or religion provide	4.3	1.28	183	39
Conformity	Restraining actions, inclinations and impulses likely to upset or harm others and violate social expectations	4.2	1.21	181	41
Achievement	Demonstrating personal success and competence according to social standards	3.4	1.29	183	39
Hedonism	Pleasure or sensuous gratification for oneself	3.3	1.31	183	39
Power	Social status and prestige, control or dominance over people and resources	2.3	1.04	183	39

Source: Modified from Schwartz (1994)

It is clear, then, that there should be a series of relationships between the values that a volunteer may hold (the 'prosocial "volunteer" identity', Hitlin, 2007: 249) and the general motivational factors relating to volunteering. Nevertheless, this is a rather simplistic model and takes no account of the importance of other factors such as norms (Bardi and Scwartz, 2003) or socio-structural factors such as gender, age, family status or occupation (Hitlin and Piliavin, 2004). In essence, then, it is argued that if values are motivational constructs (and have been used to examine environmental concern), significant associations should exist between values and general motivational factors.

It is now necessary to examine the two schemes to be used in this study to explore the links between socio-demographic factors, values and motivations. The first of these relates to values as developed by Schwartz (1994) and the second relates to functional motivations for volunteering as developed by Clary *et al.* (1998).

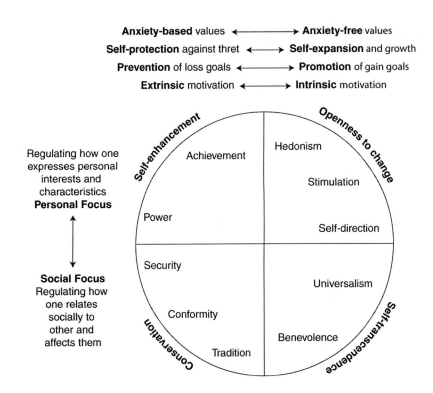

Figure 10.1 Schwartz's structure of values

Source: Schwartz, 2008

Schwartz (1994) suggests a values hierarchy in which there are 56 first order value items that are combined into ten second order values (Table 10.2). These second order values are then used to determine four third order values that can be resolved into two bipolar dimensions: 'self-enhancement-self-transcendence' and 'conservation-openness to change' (Figure 10.1). As pointed out by Schwartz (1994: 25)

> The partitioning of single values into value types represents conceptually convenient decisions about where one fuzzy set ends and another begins in the circular structure. The motivational differences between value types are continuous rather than discrete, with more overlap in meaning near the boundaries of adjacent value types. Consequently, in empirical studies, values from adjacent types may intermix rather than emerge in clearly distinct regions. In contrast, values and value types that express opposing motivations should be discriminated clearly from one another.

This circumplex structure indicates that opposing values are directly opposite each other in the model while those that are more compatible are closer to each other in the circle. Thus the higher order values of self-transcendence, concerned with the welfare of others and the environment, directly oppose those of self-enhancement, concerned with an individual's own success. Similarly, openness to change, representing change and independence, directly opposes conservation, which is concerned with stability and tradition.

Schwartz's (1994) values have been related to environmental values/attitudes. Schultz and Zelezny (1999), for example, found that universalism was positively correlated and power and tradition negatively correlated to the New Environmental Paradigm, and that ecocentric attitudes correlated with universalism while anthropocentric attitudes correlated with power. Overall, self-transcendent values were those most closely related to the New Environmental Paradigm. Aoyagi-Usui *et al.* (2003), however, found differences between Western and Asian countries; the former showing links between altruistic values and environmental values, while the latter also showed a relationship with traditional values. Values have also been linked to the leadership characteristics of environmental organisations (Egri and Herman, 2000). It was found, for example, that leaders of not-for-profit organisations scored higher on self-transcendence values than leaders of for-profit organisations who scored higher than a Canada/USA group of managers. Additionally it was shown that for not-for-profit leaders, the most important values were self-direction, universalism and benevolence. It is also worth noting that Byrne and Bradley (2007) have shown that cultural level values appear to be more important in mediating leadership styles than personal values.

The work of Clary *et al.* (1998: 1516) exemplifies the functional approach to analysing the motivations for volunteering, in which they suggest that:

> The core propositions of a functional analysis of volunteerism are that acts of volunteerism that appear to be quite similar on the surface may reflect

markedly different underlying motivational processes and that the functions served by volunteerism manifest themselves in the unfolding dynamics of this form of helpfulness, influencing critical events associated with the initiation and maintenance of voluntary helping behaviour.

As such, the functional approach offers the opportunity to explain many types of volunteering, including volunteer tourism, and particularly the reasons for becoming and remaining a National Trust working holiday leader. In addition, it meets the need to use a standard and tested scale (Petriwskyj and Warburton, 2007).

Clary *et al.* (1998) developed and tested a Volunteer Functions Inventory (VFI) that identified six generic functional motivations for volunteering derived from 30 individual motivational questions (Table 10.3). The VFI has subsequently been successfully used in a number of studies in a wide range of contexts: for example, Bradford and Israel (2004) in a study of sea turtle conservation in Florida; Finkelstein (2007) to predict satisfaction in older volunteers; Papadakis *et al.* (2004) in a study of college students; Houle *et al.* (2005) in relation to predicting task preference; Silverberg *et al.* (2000) in relation to understanding parks and recreation volunteers; and Yoshioka *et al.* (2007) in an examination of volunteers and non-volunteers. Values and understanding are frequently found to be the highest scores for volunteers whereas the protective and social factors appear lower down on the list with career being midway (Clary *et al.*, 1998; Allison *et al.*, 2002; Papadakis *et al.*, 2004). Of particular interest for environmental volunteering is the similar result found by Bradford and Israel (2004) for turtle conservation in Florida. After a concern for turtles came a concern for people and understanding, with self-enhancement and career filling the bottom places.

Based on the above three areas (socio-demographic factors, basic human values and functional motivations) a simple model is suggested in which values may be related to a variety of antecedent factors. Hitlin and Piliavin (2004) list a wide range of antecedents: biology, race/ethnicity/gender and social structure (including social class/occupation/education, family characteristics, immigrant status, age cohort, religion and national/demographic). It is further argued that values arise through the socialisation of an individual based on their socio-demographic status through various stages in their life cycle and through their position within the economic structure of society. Further, if values are seen as the foundations upon which individuals and society operate, since they are seen to be universal motivating constructs, values ultimately lead to motivations for undertaking specific actions. In the present paper, the simple model shown in Figure 10.2 is evaluated.

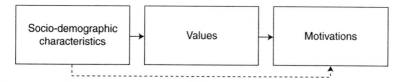

Figure 10.2 Simple model to be evaluated

Table 10.3 Dimensions of the Volunteer Function Inventory and summary statistics for volunteer leaders in the present study

Function	Conceptual definition and sample VFI items	Mean	St. dev.	N	Missing
Understanding	The volunteer is seeking to learn more about the world or exercise skills that are often unused. 'Volunteering lets me learn things through direct, hands on experience' 'Volunteering allows me to gain a new perspective on things'	4.6	1.00	200	22
Values	The individual volunteers in order to express or act on important values like humanitarianism. 'I feel it is important to help others' 'I can do something for a cause that is important to me'	4.2	1.01	201	21
Enhancement	One can grow and develop psychologically through volunteer activities. 'Volunteering increases my self-esteem' 'Volunteering makes me feel needed'	3.7	1.14	200	22
Career	The volunteer has the goal of gaining career-related experience through volunteering. 'Volunteering will look good on my CV' 'Volunteering can help me get a foot in the door at a place where I would like to work'	3.0	1.27	200	22
Protective	The individual uses volunteering to reduce negative feelings such as guilt, or to address personal problems. 'By volunteering I feel less lonely' 'Volunteering is a good escape from my own troubles'	2.9	1.11	200	22
Social	Volunteering allows an individual to strengthen his or her social relationships. 'People I'm close to want me to volunteer' 'My friends volunteer'	2.7	1.04	200	22

Sources: Clary *et al.* (1998); Clary and Snyder (1999)

Methodology

Data was collected by use of an online questionnaire. The initial questions were concerned with socio-demographic data such as age, gender, marital status, children living at home and occupation. The intention here was to utilise this data to examine the overall structure of the volunteer cohort in terms of life stages and economic status.

Schwartz's (1994) values inventory was used to collect data on values. Although this usually starts with 56 items that are then reduced to the ten in Table 10.2, the present study asked working holiday leaders directly 'How important are the following to you?' for each of the ten items using a response scale from '1. Extremely unimportant' to '6. Extremely important'.

The Volunteer Function Inventory (Clary *et al.*, 1998) was used to gather data on motivations. The use of the scale goes some way to satisfying the need to be able to make cross-group comparisons in a standard manner as suggested by Petriwskyj and Warburton (2007). For the 30 statements on the VFI, working holiday leaders were asked: 'To what extent do you agree or disagree with the following statements in relation to you as a National Trust volunteer leader?' Responses were on a scale of '1. Very strongly disagree' to '6. Very strongly agree'.

Emails were sent by the National Trust to all working holiday leaders who had agreed to be contacted in this way. The email contained a direct link to the online questionnaire hosted by SurveyMonkey. The National Trust has 823 registered leaders; the email was sent to approximately 723; 222 responded during the survey period (February 2008). This represents a response rate of just under 31 per cent, representing roughly 27 per cent of the total working holiday volunteers. Fricker and Schonlau (2002) report response rates for email surveys ranging from 6 per cent to 68 per cent with a mean of 33 per cent which compares favourably with the figure for the present study; for web-based-only surveys these authors present data that averages a 32 per cent response rate, whereas an average of 55 per cent may be calculated for mail surveys. The overall quality of the data must also be considered. Fricker and Schonlau (2002) present data from studies that compared email and postal surveys, the results suggesting that in terms of missing data there is little difference. Only two or three people in the present study did not give socio-economic data; three to fourteen did not give data on questions relating to leadership; while the questions relating to motivations showed lower response rates with up to 13 per cent of the data missing for some questions. Where possible, the largest sample size was used in describing and analysing the data to avoid data loss. As a consequence, there are small variations between tables. It has also been shown that web-based surveys may give significantly different results to telephone surveys (Roster *et al.*, 2004) but the data may be more reliable (Braunsberger *et al.*, 2007).

Initial data analysis assessed the overall structure of Schwartz's (1994) values system using multidimensional scaling. This indicated that the data in the present study conformed reasonably well to that expected. Subsequently, mean scores on the self-enhancement, self-transcendence, conservation and openness to change dimensions were calculated using the value types indicated on Figure 10.1. Similarly, the data for the VFI of Clary *et al.* (1998) was factor analysed, the results indicating a generally good, but not identical, factor structure to the original work. Subsequently, mean scores were calculated for the six motivational factors previously identified using the individual statements in the original study by Clary *et al.* (1998).

Descriptive results are presented for the socio-demographic characteristics of the leaders along with means for values and motivations. To assess the effect of socio-demographic characteristics on values and motivations, univariate ANOVA was undertaken and eta used to assess effect size of significant effects ($p<0.05$).

Partial correlation, controlling for overall value and motivation means to reduce the differential use of the scales, was used to assess the relationships between values and motivations. The results are presented in Figure 10.3, where only significant (p<0.05), positive correlations are shown (significant, negative relationships generally indicate the reverse of those shown) and etas are shown where they are associated with significant effects. All analyses were undertaken using SPSS 16.

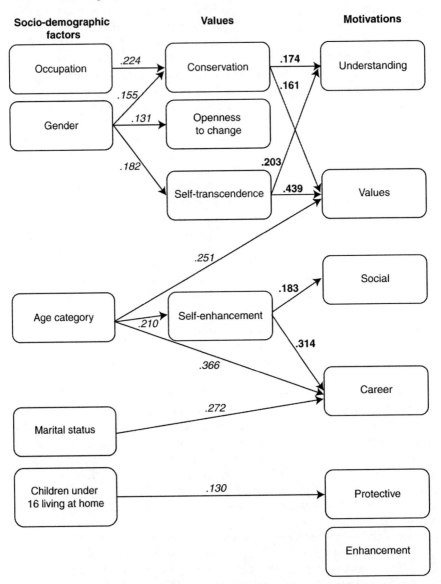

Figures in italics are etas. Figures in bold are partial correlations. All identified links are significant to at least the 0.05 level.

Figure 10.3 Interrelationships between socio-demographic factors, values and motivation

Results and discussion

Socio-demographic characteristics

The majority of volunteer leaders fall in the 30–59 age group, show a tendency to be unattached to a spouse or partner, do not have children living at home, are equally likely to be male or female and come predominately from a managerial or professional occupation (Table 10.4). These findings may be compared with previous research discussed earlier.

Table 10.4 Socio-demographic characteristics of volunteer leaders

Gender	Frequency	Percentage
Male	105	48
Female	114	52
Total	219	100

Age categories	Frequency	Percentage
20–29	22	10
30–39	72	33
40–49	54	25
50–59	42	19
60–69	27	12
70–79	3	1
Total	220	100

Marital status	Frequency	Percentage
Single	98	45
Married or living with partner	83	38
Separated, divorced or widowed	38	17
Total	219	100

Children living at home	Frequency	Percentage
No	203	92
Yes	17	8
Total	220	100

Occupational group	Frequency	Percentage
Other (inc. students)	27	12
Manual and Service Occupations	11	5
Clerical and Technical Occupations	20	9
Managerial Occupations	42	19
Professional Occupations	120	55
Total	220	100

An important difference to most of these studies is the fact that the younger age contingent is largely absent from the NT sample of leaders. This is obviously understandable in that it is likely that leaders begin as volunteers at a younger age and later progress onto leading. Additionally, many younger volunteers may not have the range of skills required to undertake the leadership role (Table 10.1). It is also clear that there is considerable similarity in the coincidence of the older age groups with previous research. In addition, there is some similarity in the preponderance of volunteers in the managerial and professional occupations or those likely to enter such occupations after completion of university studies, both groups, perhaps, having greater time available for volunteering. People in lower occupational categories tend to lead more holidays per year, but the total number of holidays led by people in professional categories is roughly the same as all other categories combined. This may indicate that such holidays are perceived as a cheaper option and those volunteers in managerial and professional occupations have sufficient financial resources to participate in other types of holiday. Those volunteer leaders that are not attached to a spouse or partner tend to lead more holidays on average per year, although not significantly so, as do those without children, indicating, perhaps, that these volunteer leaders have time available to do so. The results do not support the generally held view that females form a high proportion of volunteer tourists. The above view of NT leaders suggests they are not 'extreme, extravagant and elite' (Weiler and Richins, 1995). There appears, however, to be a distinct group of people who lead these holidays and the NT may do well, perhaps, to address other markets to broaden the leadership base.

Basic human values

The second order value scores (Table 10.2) indicate an expected pattern for volunteers with values related more closely to humanitarian or altruistic concepts towards the top of the list (universalism, security, benevolence) while those that indicate personal advancement, the more egocentric values, are towards the bottom of the list (e.g. power, achievement, hedonism). With the exception of security, those in the top half of the list represent intrinsic motivational values (Schwartz, 2008: Figure 10.1). Those in the bottom half of the list (with the exception of hedonism) represent extrinsic motivational factors. Since much of the work undertaken on working holidays is concerned with environmental management and the role of leader is predominantly looking after the welfare of the volunteers (Table 10.1), this pattern may be typical of people taking on this role. In addition, self-direction may be considered a desirable quality of leaders. The predominant values, however, are likely to depend on the situational context (Table 10.5). Leaders of organisations are likely to be motivated by self-direction values and by those that may demonstrate achievement. The latter, however, is not a priority of NT leaders suggesting to an extent that the leadership role is taken on at least in part for altruistic reasons. The position of hedonism in the second example in Table 10.5 demonstrates that for some groups this is of considerable importance, in this case a younger student group. It would seem, however, that

NT leaders are not so concerned with 'pleasure or sensuous gratification for one-self'! Power does not feature as important values for many groups (Table 10.5) while tradition is mid-table for NT leaders suggesting that they have very traditional values. This interpretation of the data is further confirmed by the third order values scores (Table 10.6) in which it is clear that self-enhancement is of low priority whereas self-transcendence and conservation are far more important motivational values. Indeed, some authors have equated Schwartz's 'conservation' dimension with traditional values (Stern and Dietz, 1994). The ranking of values is important as Schultz and Zelezny (1999), for example, have shown that universalism is positively related to pro-environmental attitudes in terms of the New Environmental Paradigm and with a measure of ecocentrism, whereas power is negatively associated with these attitudes and positively with anthropocentrism. The relevance of this to environmental working holidays should be clear. In addition, although there is no substantive evidence in the present study, there are indications (together with the roles given in Table 10.1) that leadership is generally of a transformational nature.

The overall results at this level suggest that working holiday volunteer leaders are not concerned with social status or its demonstration, but are more concerned with the development of people and the environment, while at the same time demonstrating values that may be construed as those pertaining to leadership qualities such as independent thought and challenge.

Schwartz (2008) notes several relationships between age, gender and values, but only a limited range is apparent in the present study. Schwartz (2008) suggests that values of self-enhancement (power and achievement) become less important with age, which is the case here, as does hedonism, although the effect of the latter is obscured when incorporated into the openness to change dimension. Current NT leaders have not yet reached the stage, however, where stimulation and self-direction have become less important. Gender appears to significantly affect the third-order values of conservation, openness to change and self-transcendence with females scoring slightly higher on these value dimensions. Although Schwartz (2008) suggests that this is the case for conservation and self-transcendence, he suggests that males should have higher scores than females on openness to change and self-enhancement. There is no influence of gender on the latter in the present study and it may be suggested that females taking on the role of leader need to demonstrate high values of stimulation and self-direction by taking on the responsibilities of the leader role. Life stage and children at home do not appear to have any effect on third-order values but the lack of ties seems to be an enabling factor in allowing people to take on the role of volunteer leader.

Table 10.5 Comparison of values position with previous research

Position	Leaders of for-profit organisations [a]	Leaders of nonprofit organisations [a]	Managers Canada/USA [a]	Canada [b]	Spain [b]	USA [b]	National Trust working holiday leaders
1st	Self-direction	Self-direction	Self-direction	Benevolence	Hedonism	Benevolence	Universalism
2nd	Achievement	Achievement	Achievement	Hedonism	Self-direction	Hedonism	Self-direction
9th	Power	Tradition	Power	Tradition	Tradition	Tradition	Hedonism
10th	Tradition	Power	Tradition	Power	Power	Power	Power

Notes
a Egri and Herman 2000 (managers of organisations)
b Schultz and Zelezny 1999 (undergraduate students).

Table 10.6 Schwartz's third order values of volunteer leaders in the present study

Value	Mean	Std. Dev.	N	Missing
Self-transcendence (Benevolence/Universalism)	4.7	0.92	184	38
Conservation (Security/Conformity/Tradition)	4.4	0.92	183	39
Openness to change (Hedonism/Stimulation/Self-direction)	4.2	0.94	183	39
Self-Enhancement (Power/Achievement)	2.9	0.98	184	38

Functional volunteer motivation

The functional motivations of NT volunteer leaders (Table 10.3) are very similar to other volunteers (Table 10.7). Much of the previous research has shown that values and understanding are prime motivational factors expressing a desire to demonstrate humanitarian concerns while at the same time taking an active part in the process. For NT leaders this translates into a concern for the environment that is enacted through physical activity and additionally taking the active role of leader. Bradford and Israel (2004) found similar results in a study of turtle conservation in Florida. The enhancement factor representing self-esteem is also of importance and helps the volunteers feel needed. It is noted, however, that peer pressure from friends and relatives (social factor) is of limited importance and that volunteers do not appear to be 'running away' from their own problems. Career factors often appear in the middle or bottom of the list of motives. These motivations are important because, among other reasons, they may predict task preference (Houle *et al.*, 2005) or levels of satisfaction with the volunteer experience (Finkelstein, 2007).

Relationships between socio-demographics, values and motivation

Space does not permit the presentation of the full statistical analyses; rather, significant effects ($p<0.05$) of the influence of socio-demographic factors on values and motivations are indicated in the form of etas, while significant ($p<0.05$) positive, partial correlations are given for the relationships between values and motivations (Figure 10.3).

Table 10.7 Comparison of functional motivations with the present study

Clary et al. 1998 Study 1	Clary et al. 1998 Study 2	Clary et al. 1998 Study 5	Papadakis et al. 2004	Allison et al. 2002	National Trust working holiday leaders
Values	Values	Values	Values	Values	Understanding
Understanding	Understanding	Enhancement	Understanding	Understanding	Values
Enhancement	Enhancement	Understanding	Career	Enhancement	Enhancement
Career	Career	Protective	Enhancement	Protective	Career
Protective	Protective	Social	Social	Social	Protective
Social	Social	Career	Protective	Career	Social
Varied organisations and ages	University students	Older volunteers, community hospital	University student volunteers	Non-profit organisation, Texas	This study

The impression given by Figure 10.3 suggests a complex set of linkages between socio-demographic factors, values and motivations. On closer inspection, however, two main sub-systems seem to be apparent. The first is based particularly on gender, but also occupation, operating through an influence on the values of conservation and self-transcendence to influence the volunteer motives of understanding and values. Thus, gender and occupation are directly related to a number of Schwartz's values (conservation, openness to change, self-transcendence) which in turn are directly related to Clary's volunteer motivations of understanding and values. Females have higher levels of self-transcendence, conservation and openness to change values than do males but there is no simple relationship between occupation and conservation. The link between Schwartz's self-transcendence and Clary's value motivation factor is particularly strong. This is not surprising since the former is based on universalism and benevolence while the latter is a motivation to express these values, particularly those of humanitarianism.

The second main sub-system shows the linkages between age category, the values of self-enhancement and the motivations of social and career. Age, however, may directly affect the career motivation (and is also linked to values in the sub-system discussed above). Marital status may also influence career motivation but is also related to age. Both the values of self-enhancement and the motivational factor of career show a general decline with age, whereas Clary's motivational values factor increases with age.

One other association appears on the diagram between having children at home and the protective motivational factor, with those having children at home having a lower score on the latter.

None of the socio-demographic measures or any of Schwartz's values is positively related to the motivational factor of enhancement.

The interacting system of socio-demographic variables, values and motives described above goes some way towards establishing the type of model envisaged by Lockstone *et al.* (2002) in which values are an important starting position in determining the volunteer outcomes. The system also indicates that some values relate to some motivational factors (Hitlin, 2003) but values fail to account for reinforcement of protective and enhancement motives, the latter being third most important to NT working holiday leaders. This is somewhat surprising since the value of openness to change, at least on the face of it since it contains self-direction and stimulation, would appear to be a precursor to the enhancement motivation. Similarly, the protective motivation is not predicted by any of the value dimensions although this is not an important factor for NT leaders. This is interesting since in other studies of volunteer tourism this psychological dimension is rarely if at all noted. However, Schwartz's system of values contains little that indicates this rather introspective motivation. The results presented here do not support Bardi and Schwartz's (2003) assertion that there is a single system of values and motives but rather that values could help us in understanding the motives for volunteering (Dekker and Halman, 2003). Further, since different values appear to be influencing different motivational factors, then it may be possible to segment volunteers on a values basis (Wymer, 1997).

Conclusion and implications

Given the overall findings from this study, the simple model hypothesised earlier is largely supported. Thus, there is an integrated system of socio-demographic variables, values and motives for the NT working holiday volunteer leaders in this study. What practical applications does this have? Initially, it may be pointed out that these volunteers form a distinct group and it is suggested that the NT may wish to widen the spectrum of people who volunteer to become leaders. How can this be achieved? Both values and motivations offer a means to attract potential leaders by targeting those that appeal to certain socio-demographic groups (Wymer, 1997; Clary *et al.*, 1998; Yoshioka *et al.*, 2007). This presupposes that such groups exist and this, together with further examination of the model as a whole using structural equation modelling, would provide further extensions to the work.

Acknowledgements

I am grateful to Jennie Owen and the volunteer holiday team at the National Trust for facilitating this study.

References

Allison, L. D., Okun, M. A.and Dutridge, K. S. (2002) Assessing volunteer motives: a comparison of an open-ended probe and Likert rating scales. *Journal of Community and Applied Social Psychology*, 12: 243–255.

Aoyagi-Usui, M., Vinken, H. and Kuribayashi, A. (2003) Pro-environmental attitudes and behaviors: an international comparison. *Human Ecology Review*, 10(1): 23–31.

Bardi, A. and Schwartz, S. H. (2003) Values and behavior: strength and structure of relations. *Personality and Social Psychology Bulletin*, 29(10): 1207–1220.

Bradford, B. M. and Israel, G. D. (2004) *Evaluating Volunteer Motivation for Sea Turtle Conservation in Florida*. IFAS Extension, University of Florida, AEC 372. Available at: http://www.google.co.uk/url?sa=t&source=web&cd=1&ved=0CB4Q FjAA&url=http%3A%2F%2Fredlitoral.googlepages.com%2F7EvaluatingVolunte erMotivationSeaTur.pdf&ei=Da-YTJbACJWT4garsY1U&usg=AFQjCNEPT6JjA ebdFTNKMWo8P3jvSr8jMw. Accessed 21 September 2009.

Braunsberger, K., Wybenga, H. and Gates, R. (2007) A comparison of reliability between telephone and web-based surveys. *Journal of Business Research*, 60: 758–764.

Broad, S. (2003) Living the Thai life – a case study of volunteer tourism at the Gibbon Rehabilitation Project, Thailand. *Tourism Recreation Research*, 28(3): 63–72.

Brown, S. and Lehto, X. (2005) Travelling with a purpose: understanding the motives and benefits of volunteer vacationers. *Current Issues in Tourism*, 8(6): 479–496.

Byrne, G. J. and Bradley, F. (2007). Culture's influence on leadership efficiency: how personal and national cultures affect leadership style. *Journal of Business Research*, 60: 168–175.

Campbell, L. M. and Smith, C. (2006) What makes them pay? Values of volunteer tourists working for sea turtle conservation. *Environmental Management*, 38(1): 84–98.

Clary, E. G. and Snyder, M. (1999). The motivations to volunteer: theoretical and practical considerations. *Current Directions in Psychological Science*, 8(5): 156–159.

Clary, E. G., Snyder, M., Ridge, R. D., Copeland, J., Stukas, A. A., Haugen, J. and Miene, P. (1998) Understanding and assessing the motivations of volunteers: a functional approach. *Journal of Personality and Social Psychology*, 74(6): 1516–1530.

Clifton, J. and Benson, A. (2006). Planning for sustainable ecotourism: the case for research ecotourism in developing country destinations. *Journal of Sustainable Tourism*, 14(3): 238–254.

Coghlan, A. (2008) Exploring the role of expedition staff in volunteer tourism. *International Journal of Tourism Research*, 10: 183–191.

Dekker, A. and Halman, L. (2003) Volunteering and values: an introduction. In: P. Dekker and L. Halman (eds) *The Values of Volunteering: Cross-cultural Perspectives*. New York: Kluwer Academic/Plenum Publishers, pp. 1–17.

Dietz, T., Fitzgerald, A. and Shwom, R. (2005). Environmental values. *Annual Review of Environment and Resources*, 30, 335–72.

Donald, B. J. (1997) Fostering volunteerism in an environmental stewardship group: a report on the Task Force to Bring Back the Don, Toronto, Canada. *Journal of Environmental Planning and Management*, 40(4): 483–505.

Egri, C. P. and Herman, S. (2000) Leadership in the North American environmental sector: values, leadership styles, and contexts of environmental leaders and their organizations. *Academy of Management Journal*, 43(4): 571–604.

Finkelstein, M. A. (2007). Correlates of satisfaction in older volunteers: a motivational perspective. *International Journal of Volunteer Administration*, 24(5): 6–12.

Fricker, R. D. and Schonlau, M. (2002) Advantages and disadvantages of internet research surveys: evidence from the literature. *Field Studies*, 14(4): 347–367.

Galley, G. and Clifton, J. (2004) The motivational and demographic characteristics of research ecotourists: Operation Wallacea volunteers in south-east Sulawesi, Indonesia. *Journal of Ecotourism*, 3(1): 69–82.

Halpenny, E. A. and Caissie, L. T. (2003) Volunteering on nature conservation projects: volunteer experience, attitudes and values. *Tourism Recreation Research*, 28(3): 25–33.

Hitlin, S. (2003) Values as the core of personal identity: drawing links between two theories of self. *Social Psychology Quarterly*, 66(2): 118–137.

Hitlin, S. (2007) Doing good, feeling good: values and the self's moral center. *Journal of Positive Psychology*, 2(4): 249–259.

Hitlin, S. and Piliavin, J. A. (2004) Values: reviving a dormant concept. *Annual Review of Sociology*, 30: 359–393.

Houle, B. J., Sagarin, B. J. and Kaplan, M. F. (2005) A functional approach to volunteerism: do volunteer motives predict task preference? *Basic and Applied Social Psychology*, 27(4): 337–344.

Jackson, S. (2007) Attitudes towards the environment and ecotourism of stakeholders in the UK tourism industry with particular reference to ornithological tour operators. *Journal of Ecotourism*, 6(1): 34–66.

Jackson, S. (2009) Volunteer tourism: National Trust working holiday volunteer leaders – who does it and why? Paper presented at *An international conference: tourist experiences: meanings, motivations and behaviours. University of Central Lancashire, April 1–4 2009. Conference keynote speaker paper and abstracts.* Preston: University of Central Lancashire.

Lockstone, L., Jago, L. and Deery, M. (2002) The propensity to volunteer: the development of a conceptual model. *Journal of Hospitality and Tourism Management*, 9(2): 121–133.

Manfredo, M. J. and Dayer, A. A. (2004) Concepts for exploring the social aspects of human-wildlife conflict in a global context. *Human Dimensions of Wildlife,* 9: 317–328.

Papadakis, K., Griffin, T. and Frater, J. (2004) Understanding volunteers' motives. *Proceedings of the 2004 Northeastern Recreation Research Symposium,* GTR-NE-326, 321–326.

Petriwskyj, A. M. and Warburton, J. (2007) Motivations and barriers to volunteering by seniors: a critical review of the literature. *International Journal of Volunteer Administration*, 24(6): 3–16.

Randle, M. and Dolnicar, S. (2006) *Who Donates Time to the Benefit of the Environment and Animal Rights? Profiling Volunteers from an International Perspective.* Faculty of Commerce – Papers, University of Wollongong. Available at: http://ro.uow.edu.au/commpapers/89. Accessed 28 April 2009.

Rokeach, M. (1973) *The Nature of Human Values.* New York: Free Press.

Roster, C. A., Rogers, R. D., Albaum, G. and Klein, D. (2004) A comparison of response characteristics from web and telephone surveys. *International Journal of Market Research*, 46(3): 359–373.

Schultz, P. W. and Zelezny, L. (1999) Values as predictors of environmental attitudes: evidence for consistency across 14 countries. *Journal of Environmental Psychology*, 19: 255–265.

Schwartz, S. H. (1994) Are there universal aspects in the structure and contents of human values? *Journal of Social Issues*, 50(4): 19–45.

Schwartz, S. H. (2006) *Basic Human Values: An Overview.* Available at: www.yourmorals.org/schwartz.2006.basic%20human%20values.pdf. Accessed 3 March 2009.

Schwartz, S. H. (2008) Basic values: how they motivate and inhibit prosocial behavior. Herzliya Symposium on Personality and Social Psychology. *Prosocial Motives, Emotions, and Behavior.* Interdisciplinary Center, Herzliya, Israel, 24–27 March 2008. Available at: http://portal.idc.ac.il/en/Symposium/HerzliyaSymposium/Documents/dcSchwartz.pdf. Accessed 3 March 2009.

Silverberg, K. E., Backman, S. J. and Backman, K. F. (2000) Understanding parks and recreation volunteers: a functionalist perspective. *Society and Leisure*, 23(2): 453–475.

Stern, P. C. and Dietz, T. (1994) The value basis of environmental concern. *Journal of Social Issues*, 50(3): 65–84.

Stoddart, H. and Rogerson, C. M. (2004) Volunteer tourism: the case for Habitat for Humanity South Africa. *GeoJournal*, 60: 311–318.

Vaske, J. J. and Donnelly, M. P. (1999) A value-attitude-behaviour model predicting wildland preservation voting intentions. *Society and Natural Resources*, 12: 523–537.

Weiler, B. and Richins, H. (1995) Extreme, extravagant and elite: a profile of ecotourists on Earthwatch Expeditions. *Tourism Recreation Research*, 20(1): 29–36.

Wu, H. C. (2002) Exploring the relationships between motivation and job satisfaction of volunteer interpreters: a case study from Taiwan's national park. *Proceedings of the IUCN/WCPA-EA-4 Taipei Conference* (pp.607–630), 18–23 March 2002, Taipei, Taiwan.

Wymer, W. W. (1997) Segmenting volunteers values, self-esteem, empathy, and facilitation as determinant variables. *Journal of Nonprofit and Public Sector Marketing*, 5(2): 3–28.

Yoshioka, C. F., Brown, W. A. and Ashcraft, R. F. (2007) A functional approach to senior volunteer and non-volunteer motivations. *International Journal of Volunteer Administration*, 24(5): 31–43.

11 Volunteer archaeological tourism

An overview

Jamie Kaminski, David B. Arnold
and Angela M. Benson

Introduction

Archaeology is the study of human cultures through the recovery, recording and analysis of material remains. The principal tool that archaeologists use to recover this evidence is excavation. Long before volunteer tourism became popular in the fields of conservation and the environment, development and education, volunteers have been an integral part of archaeological excavation. Originally these volunteers were linked to academic institutions. However, since the 1960s this pool has been supplemented by those with no archaeological background.

Volunteer archaeological tourism has since become increasingly popular for volunteer tourism consumers and a necessity for many excavations. Given the heavy reliance that archaeology places on volunteers for research excavation, the understanding of the scale of volunteer involvement and their contribution is relatively poor. While the study of such activity in the eco-tourism and associated fields has spawned a rich literature, archaeology has almost nothing, despite its greater antiquity. Consequently, this paper sets the framework for further discussion in respect of volunteer archaeological tourism.

Historical background

Archaeological excavation requires physical labour. This can include site clearance, the removal of topsoil, excavation of archaeological strata, the movement of spoil[1] and the backfilling of excavations on completion. Although mechanical means can be brought to bear on some of these tasks the labour required to excavate archaeological strata cannot be significantly mitigated.[2] It has to be done manually, which makes excavation highly resource-intensive.

The supply and funding of labour for excavation has been a critical but overlooked aspect of the discipline. For early modern archaeologists, the cost of labour was a significant component of the excavation budget.[3] During the history of archaeology the labour used on excavations has been dictated by funding and developments in academia. Archaeology in the eighteenth and nineteenth centuries was the prerogative of the rich. The field was dominated by antiquarians and 'gentleman amateurs' such as Sir Richard Colt Hoare (1758–1838) who excavated at Stonehenge and dug no fewer than 379 burial mounds (barrows) on Salisbury Plain. During the same period in America the politician and third president of the United States Thomas Jefferson (1743–1826) excavated the Native American burial mounds on his land.[4] Others include Heinrich Schliemann (1822–1890), a German businessman who made his name in the field of Mycenaean archaeology, excavating Hissarlik in Turkey, the site of Homer's Troy. Or George Edward Stanhope Molyneux Herbert, 5th Earl of Carnarvon (1866–1923), who funded excavations in the Valley of the Kings in Egypt, most famously leading to the discovery of the tomb of the New Kingdom Pharaoh Tutankhamun. It was these sorts of men who had both the financial ability to hire diggers and the leisure time to support their interest in archaeological excavation.

Prior to the twentieth century excavations did not involve volunteers in the modern sense. Labourers were paid for their services to dig (and find artefacts)[5] and were more or less loosely supervised by trained archaeologists. For example, Sir William Matthew Flinders Petrie (1853–1942)[6] categorised his labourers as trench diggers, shaft sinkers and stone cleaners, who were supported by earth carriers and general labourers, all of whom were supervised by Petrie. But as archaeology became more and more scientific from the 1930s onwards, less reliance was placed on untrained labourers to move earth. Increasingly, professional archaeologists trained in university archaeology departments were being employed. Increasing numbers of academic institutions across Europe had been offering archaeology as a course since the mid-nineteenth century. The graduates from these courses would provide many of the professional archaeologists who would run excavations, while the students would provide a welcome source of semi-qualified labour as part of their studies. But the use of professionals increased the cost of excavation.

Another factor that came into play was the change in funding sources for excavation. By the end of the first quarter of the twentieth century, the age of the individual funders of excavations was coming to an end. Excavations were increasingly being run under the *aegis* of academic institutions supplemented by both public and private organisations. Funding was coming from a progressively large spectrum of sources. At this time volunteer archaeological placements were oriented towards those who had a direct link with archaeology, such as archaeology students wishing to gain experience in different archaeological settings.

Things began to change in the 1960s. One of the earliest examples of volunteer archaeological tourism in the modern sense was instigated in 1961 in Israel by Yigael Yadin (1917–1984). Yadin invited students and interested amateurs from North America and Europe to join his expedition to the Nahal Hever cave in Judea.[7] The significance of the Nahal Hever Cave excavation was that Yadin was seeking foreign volunteers with no archaeological background to participate on an excavation. Importantly, this coincided with increased opportunities for international travel for potential tourists. The Nahal Hever Cave was a relatively small excavation. It would be two years later that Yadin would come to excavate one of Israel's most iconic sites, the hilltop palace and fortress at Masada. It was here that volunteer archaeological tourism really took off.

Overlooking the Dead Sea, on the eastern edge of the Judean Desert, Masada was occupied from at least the Bronze Age. It was the biblical Herod the Great's Palace and fortress, but, most crucially for historians and Israelis, it was the site of the siege immortalised by the Jewish historian Flavius Josephus in his *Bellum Judicum*.[8]

Yadin could apply the lessons learned from the use of volunteers at the Nahal Hever Cave to one of Israel's great archaeological sites. In 1963 Yadin advertised for volunteers in Israeli papers and in the *Observer* newspaper in the United Kingdom. As with the Nahal Hever Cave excavation, no background in archaeology was required from the volunteers. The advertisement would be the blueprint for future archaeological excavations wishing to make use of volunteers. For the Masada excavation, the volunteers were required to pay their travel expenses to the site and to stay at the dig for a minimum of two weeks. The advertisement was also forthright about the potential for difficult living conditions at the camp site. However, the opportunity to excavate on such a historically famous archaeological site, with deep meaning for Jewish national identity, brought volunteers from across the globe. The sheer number of volunteers who signed up for the expedition allowed Yadin to run 23 two-week shifts with an average of 300 participants each (Yadin, 1966: 13–14; Bacon, 1971: 181; Atkinson, 2006: 14).

Aside from excavating on an iconic site, the organisers knew that they needed to provide additional value for the volunteers to sustain the experience. Using the experience of running academic excavations with student volunteers, the supervisory staff gave evening lectures and conducted field trips on the day off. The practice of providing lectures, guided tours and field trips in conjunction with providing academic credit for field work is the foundation for many volunteer excavations today.

Some indication of the importance of the contribution that the volunteers made to the excavations was acknowledged by Yadin when he dedicated his book about Masada to them. This success would not go unnoticed. It was plainly evident that volunteer archaeological tourism could provide the labour to sustain major excavations, and it was adopted rapidly. Eliezer Oren of Ben-Gurion University, who directed excavations near Beer-sheba in the Negev, in the south of Israel, also made use of volunteers. In an interview in 1984 he summed up the importance of volunteers in the changing funding climate for archaeology: 'without volunteers, we would not be able to conduct the excavations. We don't have the ability to hire the people. The costs are tremendous.' He also preferred volunteers to students because 'they know what hard work is, and they're very dedicated' (Rosen, 2006: 20).

In many countries the cost of running archaeological excavations has increased considerably in the last two decades. Increased costs for staff, health and safety, insurance[9] in conjunction with an increasing need for specialist services (geophysics, other scientific tests)[10] and post-excavation costs have all driven up expenditure. By reducing the requirements for a paid staff, the use of volunteers has been instrumental in funding excavations. Furthermore, many excavations now charge volunteers for attending which can in some cases provide an income stream that can supplement the excavation funds. This can benefit the excavation by supporting greater periods in the field, the use of more advanced equipment or the funds for post-excavation work.

The volunteer archaeological tourism market

Today, volunteer archaeological tourist opportunities can be found in almost any geographical environment where there is archaeology, from urban and rural sites, to deserts and even underwater (marine archaeology). As such, volunteer opportunities can be found on six continents. But it is apparent that volunteer archaeological opportunities for tourists are not evenly distributed and this distribution differs from other kinds of volunteer opportunities. The majority of non-archaeological volunteer placements are made in Central and South America, Africa and India (Wearing, 2001: 2; Callanan and Thomas, 2005: 192). In contrast volunteer archaeological tourism is concentrated in Europe, North America and to a lesser extent Asia (see Figure 11.1). Therefore, volunteer tourism takes place in both developing and developed countries, but it is dominated by the developed world.

The relative distribution of volunteer archaeological opportunities shown in Figure 11.1 can be partially explained by the distribution of recognised archaeological sites, in conjunction with the distribution of archaeological organisations and the culture of excavation.[11] It is apparent that the lowest number of volunteer opportunities is found in Oceania and Africa. In both of these continents there are simply fewer archaeological sites that are being excavated.

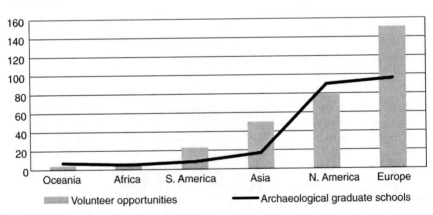

Figure 11.1 A comparison of volunteer archaeological opportunities available to tourists across continents compared with graduate schools offering archaeology in 2009

Source: Past Horizons (www.pasthorizons.com) web database (August 2009), with additions. Where a single University offers archaeology in different departments, such as anthropology and classics, this is counted as one site.

For example, prior to colonisation Australia was occupied by semi-nomadic hunter gatherers, who have left little material culture. The relative number of archaeological sites per unit area, compared to Europe for example, with its higher density occupation, is naturally less. Correspondingly there are fewer archaeology departments in Australian universities which further reduces the potential for archaeological excavation. On the African continent a similar situation of a comparatively low density of archaeological sites per unit area is compounded by areas of poverty and political instability. There are also a very low number of academic institutions with archaeology departments. Much of the active archaeology here takes place in the south and north of the continent. In both Australia and Africa there are relatively few private organisations which organise and support archaeological excavation.

The numbers of volunteer opportunities on the Asian continent are heavily skewed by the density of such excavations in the Middle East. Countries such as Israel have a highly developed archaeological framework, an extensive diaspora which is deeply interested in its history, and archaeological sites from the 'bible' lands, of interest to both Judeo-Christians and biblical scholars. In consequence Israel has the fourth-highest number of excavations with volunteer tourist opportunities after Italy, the United Kingdom and the USA (see Figure 11.2).[12]

The domination of North America and Europe in the placement of archaeological volunteer tourists is unsurprising. Archaeology as we know it today began in Europe. Europe has the greatest concentration of academic archaeology departments compared to any other continent. Furthermore, there are almost 1500 archaeological societies, foundations and other organisations and the study of the past is a cultural norm (see Figure 11.3). Europe's comparatively long history of dense occupation has led to a plethora of archaeological sites.

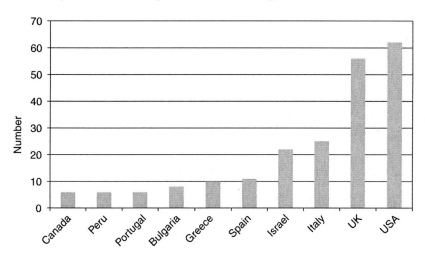

Figure 11.2 A comparison of the principal countries providing volunteer archaeological opportunities for tourists in 2009

Source: Archaeological Resource Guide for Europe http://odur.let.rug.nl/~arge/Countries/

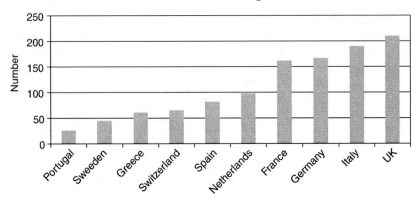

Figure 11.3 The number of archaeological groups and organisations in European countries in 2000

Source: Past Horizons (www.pasthorizons.com) web database (August 2009), with additions

This differs from North America where there are fewer known archaeological sites per unit area but culturally there is a deep desire to understand both identity and the past. There are also the resources to take part in such placements. Academic archaeology departments are also well represented. The size of the continent and the comparatively high average disposable income and available leisure time also means that many excavations can be supported by in-country volunteer tourists.

The South American continent represents a middle ground between the high density of archaeological sites in Europe and the lower densities seen in Oceania and Africa. The South American continent has remains from advanced civilizations. The northern part of Central America includes Maya, Aztec, Olmec and Toltec civilisations. The southern part of the continent includes Inca, Moche, Nasca and Tiwanaku cultures. However, there are relatively few academic departments specialising in archaeology, but the close proximity to the USA has led many North American archaeologists to conduct research projects in the south.

Archaeology and other volunteer opportunities

By the 1980s the use of volunteers in archaeology was unmatched in any other field (Rosen, 2006: 20). But although archaeology has a longer history in its use of volunteer placements, it has since been eclipsed numerically by many other forms of volunteer opportunity (see Figure 11.4). This is principally because there is a limit to the number of archaeological excavations that take place each year, and of these not all can, or are willing to, accept volunteer placements.

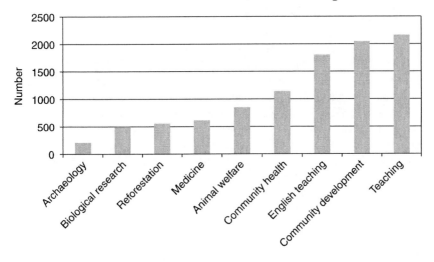

Figure 11.4 A comparison of different types of volunteer opportunities on the Projects Abroad website

Source: Projects Abroad (www.projectsabroad.com) web database (August 2009)

The limitation on the number of excavations accepting volunteers globally is determined by a number of factors:

- *There has to be archaeology.* Archaeology is the evidence of past human activity. That evidence is not evenly distributed across the globe. Some societies, such as hunter gatherers, have left far less evidence of their passing, compared to more sedentary societies. Therefore, there is a disparity in the distribution of archaeological sites.
- *Some countries are difficult to work in.* Some countries are plagued by political instability, unrest and other safety concerns. Others may have complex visa requirements for entry reducing the potential of short-term volunteer-orientated visitors (cf. Kenoyer 2008).
- *There has to be the capacity and resources to conduct excavation.* This includes financial resources, academic departments or other archaeological institutions and the culture of archaeology. This is because archaeology is a specialist technique; there are only so many university departments and archaeological organisations that can run excavations. Some countries simply do not have a well-established archaeological infrastructure and so are reliant on outside organisations conducting the majority of excavations.
- *Ability to accept volunteers.* Not all excavations have the ability or willingness to accept tourist placement volunteers. Such an undertaking may require marketing efforts in different channels. Tourist placements require administration and training, and they can carry a greater administrative overhead compared to local volunteers. For example, tourist volunteers may require

higher standards of safety and hygiene than is normally available in some countries (McCloskey, 2003: 11). Some excavations either do not have the capacity or the willingness to take on volunteers.

- *Language support.* Most consumers of volunteer archaeological tourism experiences come from the advanced economies (especially the USA and UK) but seek experiences in both the developed and developing world. Those from advanced economies are more likely to have the disposable income and free time to take part in such activities. The majority of volunteer archaeological opportunities outside of the English-speaking world need to offer support for English language in order to attract these volunteers in any great numbers. This can act as a limiting factor for the number of excavations which can accept volunteers.[13]

Opportunity providers

In this context the 'opportunity provider' is the organisation which provides the opportunity to the market. This may differ from the funder or the organiser.[14] Three groups of provider are evident:

- *The intermediaries who package the opportunity.* As increasing numbers of archaeological sites offer volunteer opportunities to potential tourists, so there is a business need for intermediaries to link the suppliers (archaeological excavations) with potential consumers of the experience. In some cases intermediaries may package the archaeological opportunity and market it to the public. However, these organisations do not run the excavations. They also operate in different locations such as Europe (Grampus Heritage),[15] or globally (Earthwatch)[16] or specific regions such as ArchaeoSpain.
- *Academic institutions.* Although dominated by university archaeology departments who have the capacity and the potential to access funds, colleges and even high schools run excavations that offer volunteer placements.
- *Other organisations.* There are numerous organisations that run archaeological excavations. These can be private organisations such as local societies, archaeological centres, trusts, research foundations, and private museums.

There are numerous online *fora* that contain listings of volunteer opportunities on various archaeological sites. These may specialise in archaeological opportunities only such as the Biblical Archaeology Society for Middle Eastern Archaeology and the archaeology of the Holy Land; others may be more generic such as Archaeologyfieldwork.com, or the Council for British Archaeology. Some may include a variety of projects including archaeology such as Projects Abroad.[17]

The comparison shown in Figure 11.5 provides a guide to the relative contribution of the different opportunity providers in different continents.

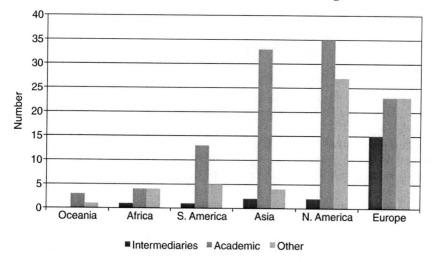

Figure 11.5 A comparison of volunteer archaeology opportunity providers on different continents

Source: Past Horizons (http://www.pasthorizons.com) web database (August 2009), with additions

Europe has a well-established archaeological infrastructure and this manifests itself in the comparative balance between opportunity providers. Academic and other providers provide equal numbers of opportunities, while the intermediaries only lag slightly.

In North America the opportunities are provided predominantly by academic institutions. However, there is a difference in the composition of those institutions compared to Europe. Although in North America universities still dominate the field, there are examples of colleges and even high schools that run excavations that provide volunteer placements. This contrasts to Europe where virtually no institutions below university level provide excavation opportunities for volunteers. North America also has far fewer opportunities from the intermediaries.

In Asia the opportunity providers are massively dominated by academic institutions. There are far fewer private organisations providing archaeological placements and fewer intermediaries. There are a number of reasons for this. The Middle East contains numerous archaeological sites which are attractive to academic institutions, either from countries within the region (almost exclusively Israel), and to foreign university departments (almost exclusively from the USA). In the rest of Asia (with the exception of China) there is a far less developed archaeological infrastructure, as evidenced by the dominance of excavations run by, or otherwise supported by, foreign academic departments and the almost complete lack of non-academic organisations providing volunteer opportunities.

In contrast South America is dominated by academic institutions, principally US universities. There are far fewer organisations in the 'other' category, highlighting that the archaeological infrastructure on the continent is underdeveloped.

There are also comparatively few opportunities provided by intermediaries, which contrasts with other volunteer areas such as volunteer ecotourism.

In Oceania and Africa, there are too few opportunities to make a meaningful comparison; however these opportunities are dominated by academic institutions. The absence of 'other' organisations highlights the relatively undeveloped archaeological culture and infrastructure on these continents.

The market

The market for volunteer archaeological tourism can be roughly divided into two groups: students and the serious leisure volunteer.

Students

Students have long underpinned the workforce on many archaeological excavations. Most students attend the research excavations run by their own universities. However, today fewer archaeological courses have student research excavations because of the cost of running such enterprises. With a difficult job market in archaeology, many students who wish to excel use volunteering opportunities to increase the variety of archaeological sites in their portfolio, which can be beneficial for their CVs. To cater for this numerous volunteer opportunities are oriented towards students such as the growing number of archaeological field schools which provide academic credit. Some even conduct exams and give qualifications.

Serious leisure

When Yadin removed the requirements for an archaeological background for taking part on excavations, he changed the entire perspective of the field. Since that time it has been possible for interested amateurs to take part in excavations. This increased the number of people who could take part and so reducing the reliance of archaeological excavations on students for labour. Volunteers with numerous motivations could then help at archaeological excavations (cf. Miller 2006). Both amateurs and volunteers that take part in archaeological digs can be captured within the serious leisure framework as outlined by Stebbins (1992; 2001; 2005; 2006); however, whilst the framework is well known and used extensively, e.g. in a sporting context, there is little research using the framework associated with archaeology.

Archaeological stakeholders

Archaeological excavations are run and funded by a number of different stakeholders:

- *Private funders*. Originally most archaeology was funded by wealthy individuals or institutions. However, these benefactors are no longer a major feature of the modern funding landscape. Today the role of wealthy individual funders has been superseded by private organisations such as archaeological trusts, and foundations.

- *Local societies.* In some countries such as the UK and USA there are numerous local archaeological societies. These sometimes fund research excavations. However, the funds and resources that these societies can provide are often supplemented by grants from other sources.
- *Developer funded excavation.* In some countries planning permission is contingent on an archaeological survey being conducted if there is considered to be a potential threat to archaeology.[18] This can represent the majority of archaeological activity in a country. There are very few opportunities for volunteers of any kind on developer-funded excavations. These require a highly professional staff that can work quickly and efficiently, within the confines of the necessary health and safety issues associated with working on construction sites with tight deadlines. As Aldous (1998: 33) states 'archaeology has become a very professional exercise. In urban development sites at least, it is working to such stringent specifications and timetable that it cannot afford to employ volunteers'.
- *Academic institutions.* University-led excavations have been the mainstay of much archaeology in the last century, for example, the University of Nottingham excavations at Crickley Hill,[19] UK where over 3,000 volunteers from around the world worked on the site in 25 seasons between 1969 and 1994, or the University of Reading's excavations in the Roman town of Silchester,[20] UK which have run annually since 1996.

With the exception of developer-funded excavations most archaeological excavations can provide a meaningful experience to volunteers. Most archaeological tour operators often select longer running excavations as a destination because these are a known quantity. Previous results can be used in marketing literature and the infrastructure is often better developed and is less likely to experience teething problems.

Why archaeology?

In Western society archaeology appears to hold a deep attraction for the public (Jameson, 2003: 160). This is one of the reasons that so many volunteer archaeological tourists come from the developed world.

In these countries the media has had a profound effect on public perceptions. There have been numerous films with 'archaeological' and pseudo-archaeological content that have fired the public imagination (although often with an erroneous idea of what constitutes archaeology). Furthermore, the link between the television and archaeology has a long history. In the UK archaeologists have long used the media to convey their work to the public. For example, Sir Mortimer Wheeler (1890–1976) conducted numerous popular excavations that made early use of television documentaries. Wheeler hosted three television series with archaeological content: *Animal, Vegetable, Mineral?* (1952–1960),

Buried Treasure (1954–1959) and *Chronicle* (1966). His books (for example, *Still Digging*) helped further the popular appeal of archaeology (Wheeler, 1955). This close link between the media and archaeology continues to the present day with a plethora of popular archaeology programmes such as *Time Team* (1994–) and *Meet the Ancestors* (1998–2004), which have caused a huge upsurge in public interest in the UK. It is not just television and film that has portrayed archaeology to the public. Archaeological coverage has been increasing in paper media as well. Archaeological reporting increased by 700 per cent in one 18-month period during the mid 1990s (Holtorf and Drew, 2007: 45). Volunteer archaeological tourism has undoubtedly benefited from the portrayal of archaeology in the mainstream media. Of course, the provision of volunteer archaeological opportunities represents another mechanism for increasing public support for the long-term benefit of the sector.

Volunteers at excavations

The difference between archaeology and the sciences is that once a site is excavated it cannot be re-excavated at a later date when techniques and methods improve. Archaeology is an 'unrepeatable experiment' (Barker, 1982: 12). Archaeology therefore sits uneasily with the sciences. It uses scientific techniques, but it is difficult to corroborate using 'experimental' process.

This yields a dilemma for the archaeological site director. Archaeological remains provide evidence of the development of our civilisation, but they are an irreplaceable resource that is both 'finite and non-renewable' (see PPG 16: A3 and PPG 16: A6). Will using volunteers produce as high quality a result as professional archaeologists? With archaeology there is only one chance to acquire the evidence required for interpretation. In the context of volunteer archaeologists, this places a great deal of pressure on the supervisors to ensure that the volunteers are properly trained and supervised during the excavation.

But even this may not be enough. Analysis of excavation records from the Neolithic site at Scara Brae,[21] in the Orkney Islands, UK, by Clarke (1978) appears to show that a differential recovery rate of artefacts can be expected among volunteers on an excavation. In this case the volunteers were all students with previous archaeological experience. Clarke (1978) feels that the differentials can be attributed to the individual interests of the volunteers and their background knowledge as much as the actual distribution of material in the ground. As a qualifier Clarke (1978: 69) notes that 'nothing in the above information constitutes a clear demonstration of the actuality of a wide range of recovery rates among volunteers on an excavation which does not accurately reflect the distribution of finds in the ground but there is much that points to it being so'.

Archaeological students have long learned their trade on training excavations where the methods and techniques of archaeological excavation can be taught under supervision (Davis, 1989: 277). With volunteers, supervision is the key.

Because of the potential for mistakes during the learning curve there can be a temptation to place volunteers in less sensitive and often less exciting jobs such as site clearance, movement of spoil, washing finds, etc (Kenyon, 1952: 65). Such jobs may not provide volunteers with as meaningful an experience as excavation, but they are representative of the totality of archaeology.

To many volunteers 'archaeology' is synonymous with excavation.[22] Of course, excavation is a crucial part of archaeology because it is the means of acquiring data, and is the most publically evident element of archaeology. However, even on an archaeological dig there is a greater range of activities than just excavation. These could include: recording (drawing plans and elevations), photography, finds processing, environmental sampling and surveying.

Some volunteer archaeology programmes emphasise these alternative components of archaeology. For example, in the USA, the Forest Service runs the 'Passport in Time' volunteer archaeology and historic preservation program, where volunteers can work with professional archaeologists on projects that, in addition to archaeological excavation, include rock art restoration, survey, archival research, and historic structure restoration (Osborn and Peters, 1991; Osborn, 1994; Jameson, 2003: 156). Such programs tend to be oriented more towards the local volunteers, however, rather than volunteer tourism.

Some archaeological sites are more complex to excavate than others. For example, complex stratigraphic sequences can be found on some urban sites. Conversely far more ephemeral stratigraphy can be found on sites from sites of short occupation/activity, such as those of nomadic peoples or battlefield sites. Such stratigraphic sequences can easily be missed or misread by those without proper training. None of this precludes the application of volunteer archaeologists to these sites, but in these instances volunteer archaeology can only be 'run responsibly with the supervision of many assistants'. As a consequence some sites have developed alternative volunteer experiences which place a greater emphasis on supervised laboratory work rather than excavation (Jameson, 2007: 371).

There is a great variety of potential options available to prospective volunteers, ranging from a couple of hours to an entire season. For example the 'Dig for a day' scheme run by Archaeological Seminars is designed for visitors of all ages. The program runs for only three hours and includes: digging, sifting and pottery examination at Tel Maresha, the ancestral home of King Herod in Beit Guvrin, Israel. The opportunity is aimed at providing practical experience for those who want to excavate but have limited time, mainly tourists.

With such variety in the types of opportunities, the attendance fees for volunteers at excavations vary widely. Some organisations require a contribution to the project, while others offer packages with full room and board. Travel costs are usually not included.

Some sites are run by local societies and may only require a small membership fee to attend. In some cases there is no fee; however, the volunteer is unlikely to receive any addition benefits other than taking part in the excavation itself. These types of excavation are often orientated towards local volunteers and students, and may be less attractive to tourists.

Some sites have a policy of 'progressively decreasing cost', where the cost to the volunteer deceases with the length of time that the volunteer attends the excavation. The economic reasoning behind this is that the longer the volunteer stays at a particular site the more useful they become because of the experience they have gained on that site.[23] The steep learning curve experienced by a new volunteer means that the greatest administrative burden is within the first week to two weeks of training. After this period the volunteers can contribute more with less intensive supervision. In a few rare instances some projects may offer a stipend to volunteers to cover some basic expenses, therefore increasing the attractiveness of the offer.

Some archaeological excavations are becoming increasingly sophisticated in their offerings to volunteers. Some are even using the example of the intermediaries to provide a wide spectrum of offerings. For example, the Maya Research Program offers programmes in archaeology and anthropology. By building a museum and visitor centre in the village nearest to the archaeological site it is hoped that tourists will visit and help to regenerate the local economy. Between 1992 and 2006, over a thousand students and volunteers worked on the excavations at Blue Creek (Guderjan, 2007: ix). Blue Creek is among the longest-running research projects in the Maya area, and the organisers are clear that 'this has been made possible principally through the participation of Maya Research Program volunteers'.[24]

The long-term, incremental impact of volunteers can be considerable. At George Washington's Mount Vernon[25] mansion volunteers and interns have helped the site's archaeologists since 1987 donating more than 50,000 hours towards researching and restoring the estate.[26]

Between 1992 and 2004, as part of the US Forest Service's Passport in Time program at Gifford Pinchot National Forest 339 volunteers contributed 15,389 hours of volunteer labour to heritage projects. Of these 10,262 hours were for purely archaeological projects.[27] The Gifford Pinchot National Forest is only one of 137 National Forests in the US that participates in the Passport in Time program. While the monetary benefits of this donated labour are often stressed, there are also other non-monetary benefits which are equally important including positive public relations, advocacy, education, community engagement and increased support from elected officials. The impact of volunteering in archaeology is considerably more diverse than simple labour cost savings.

Conclusions

Archaeological excavations have for a long time used volunteers. Depending on the nature of the organisational structure these could be members of the local archaeological society, interested locals, or those from farther afield. The origins of volunteer archaeological tourism can be traced back to the 1960s, a time when such opportunities were matched by increased travel opportunities. The use of volunteer tourists on archaeological excavations has exploded since this time but it is approaching its natural limits. The number of excavations that can take place at any time is constrained by the availability of archaeological sites, funding and qualified archaeologists to run them. However, without volunteers and volunteer tourists, the archaeological sector could not conduct the range of research that it does.

Despite this, the fact remains that the literature discussing volunteers in the context of archaeology remains minimal. Little is known about the scale and impact of volunteer archaeological tourism on the profession. In order for this sector to make the most of this resource, more research needs to be conducted in this area. With further research, the potential for improving management, monitoring and volunteer retention is considerable.

Notes

1 In archaeology, spoil is the term used for the soil, rubble and other debris that is generated by an excavation; it is discarded off site on spoil heaps.

2 There is an increasing use of machine excavators, especially in developer-driven excavations because of time pressures (Van Horn *et al.*, 1986); however, machines are used primarily to remove modern overburden and for the movement of spoil.

3 During the early days of modern archaeology there was less emphasis and resource placed on 'post-excavation' work.

4 Much of Jefferson's 'archaeological' work was conducted in the last two decades of the eighteenth century, prior to him becoming the president in 1801. His study of the Indian burial mounds led him to develop the concept of stratigraphy, and hence has earned him the title in some quarters of the 'father of archaeology'.

5 For example, in common with archaeological practice at the time, workers employed by Max Mallowan on his excavations at Tell Brak, Syria, were paid a fixed daily amount. But this was complemented by an extra payment, called 'bakshish' for small finds of archaeological significance that were made by the worker. This incentivised the workforce to dig carefully and therefore recover as many finds as possible (Barmby and Dolton, 2006; McCall, 2001: 109).

6 Professor Sir William Matthew Flinders Petrie was an Egyptologist and a pioneer of the systematic methodology. He held the first chair of Egyptology in the United Kingdom, the Edwards Professorship of Egyptian Archaeology and Philology at University College, London. He excavated numerous Egyptian archaeological sites such as Naukratis, Tanis, Abydos and Amarna, but was most famous for the discovery of the Merneptah Stele.

7 Nahal Hever was excavated by Yadin over two seasons in 1960 and 1961. In the 'Cave of Letters' three different document hoards were discovered; these included letters from Shimeon ben Kosiba (the so-called 'Prince of Israel'), the family archives of a woman called Babatha, and a fragment of biblical psalm scroll. These were documents owned by fugitives from the Romans during the Bar Kokhba revolt between AD 132 and 136 (Cansdale, 1997: 90–91).

8 Here in AD 66 a number of Jewish Zealots overcame the Roman garrison and used the fortress as a base for raiding Roman settlements. In AD 72 the Roman army besieged the fortress, finally overcoming the defences after constructing an immense 114m high ramp up to the walls. Upon entering Masada the Roman forces found the Zealots had committed mass suicide rather than face capture.

9 In the developed world insurers tend to classify archaeology in the same group as more dangerous professions such as construction (Connolly, 2009).

10 In the 1960s when Yadin was excavating Masada archaeological science was in its infancy. Specialist techniques such as radiocarbon dating, geophysical survey (resistivity, magnetometer, ground penetrating RADAR) were not widely used or had not yet been developed for archaeological applications. Despite their obvious benefits these techniques all impose an additional financial cost on site budgets.

11 The number of archaeological organisations, such as academic departments, government funded archaeologists, and private agencies give an indication of a country's capacity to conduct archaeological excavation. The culture of archaeology refers to a country's interest in archaeology.

12 Volunteer archaeology placements are even available on government websites.

13 English is currently the *lingua franca* for many tourism experiences (Maggi and Padurean, 2009).

14 This can be very complex with many organisations involved in excavations. For example, the Sino-American Field School of Archaeology, which was established in 1990, is a collaboration between the Fudan Museum Foundation and Museum of Asian Art, Sarasota, Florida, the Xi'an Jiaotong University and the Shaanxi Institute of Archaeology. The field school runs an annual archaeology practicum in China (McCloskey, 2003: 194).

15 Grampus Heritage and Training is a non-profit making organisation based in the North West of England that provides links to volunteer opportunities on European archaeological and heritage sites.

16 The EarthWatch Institute has its origins in ecotourism but EarthWatch Archaeological digs run in various parts of the world. The price of each project, called the Share of Cost (SOC), covers food, accommodation, on-site travel, and all of the various costs of field research (field permits, equipment, etc.).

17 http://www.projects-abroad.com

18 The Planning Policy Guidance 16 (PPG16) guidance note revolutionised field archaeology in Britain. It laid down that developers, by an extension of the 'polluter pays' principle, should pay for pre-development examination of their sites under the supervision and to the satisfaction of the local planning authority.

19 Crickley Hill is a multi-period hilltop site in Gloucestershire, UK. It was occupied intermittently from around 4000 BC. Originally it was a Neolithic causewayed enclosure that was remodelled into a defended enclosure that was destroyed during conflict in the Middle Neolithic. The hilltop was reused as a hill fort between the seventh and third centuries BC.

20 Silchester in Hampshire, UK, was an Iron Age Oppidum in the territory of the Atrebates, which became the Roman city of Calleva Atrebatum. The University of Reading has been excavating there intermittently since the 1970s.

21 Occupied between 3100–2500 BC, Scara Brae is a Neolithic stone built settlement comprising ten houses. It is one of the most complete Neolithic villages in Europe and as such has gained World Heritage Status.

22 Furthermore, archaeological work related to excavation carries on throughout the year. Organising the excavation, conducting post-excavation, analysing the results, and publishing are all activities that take place throughout the year; activities that most volunteers are blissfully unaware of.

23 Even semi-experienced volunteers will have a learning curve as they attend different archaeological excavations in different locations. Differences in the geology and soil conditions, types of finds, all require time to assimilate.

24 http://www.mayaresearchprogram.org/web-content/aboutus_history.html

25 Mount Vernon was home to George Washington for more than 45 years. First known as Little Hunting Creek Plantation, George Washington inherited the property upon the death of his brother Lawrence's widow in 1761. Over the years, Washington enlarged the residence and built up the property from 2,000 to nearly 8,000 acres. Today the mansion has been restored to its appearance in 1799, the last year of Washington's life.

26 At Mount Vernon excavations carry on throughout the year from Monday to Friday. Volunteers need to commit to at least half a day each visit. The site offers no lodgings or stipends, and cannot pay any expenses.

27 http://www.fs.fed.us/gpnf/research/heritage/pit.shtml

References

Aldous, T. (1998) Archaeology at the crossroads. Efforts to make archaeology more 'people-friendly'. *History Today*, 48(1) 33–4.

Atkinson, K. (2006) Diggers – From paid peasants to eager volunteers. In K. E. Miller (ed.) *I Volunteered For This?! Life on an Archaeological Dig*, Washington: Biblical Archaeology Society, pp.10–17.

Bacon, E. (1971) *Archaeology: Discoveries in the 1960s*, Praeger: New York.

Barker, P. (1982) *Techniques of Archaeological Excavation*, London: Batsford.

Barmby, T. and Dolton, P. (2006) *The Riddle of the Sands? Incentives and Labour Contracts on Archaeological Digs in Northern Syria in the 1930s*. University of Aberdeen discussion paper 2006–10 Aberdeen: University of Aberdeen.

Cansdale, L. (1997) *Qumran and the Essences: A Re-evaluation of the Evidence*, Mohr: Tübingen.

Callanan, M. and Thomas, S. (2005) Volunteer tourism. Deconstructing volunteer activities within a dynamic environment. In M. Novelli (ed.) *Niche Tourism: Contemporary Issues, Trends and Cases*. Oxford: Butterworth-Heinemann, pp.183–200.

Clarke, D. V. (1978) Excavation and volunteers: a cautionary tale, *World Archaeology*, 10(1): 63–70.

Connolly, D. (2009) *Insurance for Archaeology*. Luggate Burn: BAJR practical guide, British Archaeological Jobs and Resources.

Davis, H. A. (1989) Learning by doing: this is no way to treat archaeological resources. In H. Cleere (ed.) *Archaeological Heritage Management in the Modern World*. London: Unwin Hyman, pp. 275–79.

Guderjan, T. H. (2007) *The Nature of an Ancient Maya City: Resources, Interaction and Power at Blue Creek, Belize*. Tuscaloosa: The University of Alabama Press.

Holtorf, C. and Drew, Q. (2007) *Archaeology is a Brand: The Meaning of Archaeology in Contemporary Popular Culture*. Oxford: Archaeopress.

Jameson, J. H. (2003) Purveyors of the past: education and outreach as ethical imperatives in archaeology. In L. J. Zimmerman, K. D. Vitelli, and J. Hollowell-Zimmer, (2003) *Ethical Issues in Archaeology*. Walnut Creek: Altamira Press, pp. 153–162.

Jameson, J. H. (2007) Making connections through archaeology: partnering with communities and teachers in the National Park Service. In J. H. Jameson and S. Baugher (eds) *Past Meets Present: Archaeologists Partnering with Museum Curators, Teachers and Community Groups*. New York: Springer, pp. 339–366.

Kenoyer, J. M. (2008) Collaborative archaeological research in Pakistan and India: Patterns and processes, *The SAA Archaeological Record*, 8(3): 12–22.

Kenyon, K. M. (1952) *Beginning in Archaeology*. London: Phoenix House.

Maggi, R. and Padurean, L. (2009) Higher tourism education in English – where and why? *Tourism Review*, 64: 1: 48–58.

McCall, H. (2001) *The life of Max Mallowan: Archaeology and Agatha Christie*. London: The British Museum Press.

McCloskey, E. (2003) *Archaeo-volunteers: The World Guide to Archaeological and Heritage Volunteering*. Milan: Green Volunteers.

Miller, K. E. (2006) *I Volunteered For This?! Life on an Archaeological Dig*. Washington: Biblical Archaeology Society.

Osborn, J. A. (1994) Engaging the public, *CRM*, 17(6): 15. Washington: National Park Service.

Osborn, J. A. and Peters, G. (1991) Passport in time, *Federal Archaeology Report*, 4: 1–6.

Rosen, E. E. (2006) The volunteer's contribution to archaeology and vice versa. In K. E. Miller (ed.) *I Volunteered For This?! Life on an Archaeological Dig.* Washington: Biblical Archaeology Society, pp. 18–20.

Stebbins, R. A. (1992) *Amateurs, Professionals, and Serious Leisure.* Montreal and Kingston, Canada: McGill-Queen's University Press.

Stebbins, R. A. (2001) *New Directions in the Theory and Research of Serious Leisure.* Mellen Studies in Sociology, vol. 28. Lewiston, New York: Edwin Mellen.

Stebbins, R. A. (2005) Project based leisure: theoretical neglect of common use of free time, *Leisure Studies,* 24(1): 1–11.

Stebbins, R. A. (2006) *Serious Leisure: Perspective for our Time.* New Brunswick, NJ: AldineTransaction.

Van Horn, D. M., Murray, J. R. and White R. S. (1986) Some techniques for mechanical excavation in salvage archaeology, *Journal of Field Archaeology,* 13(2): 239–244.

Wearing, S. (2001) *Volunteer Tourism: Experiences that Make a Difference.* New York: CABI Publishing.

Wheeler, M. (1955) *Still Digging.* London: Michael Joseph.

Yadin, Y. (1966) *Masada: Herod's Fortress and the Zealot's Last Stand.* New York: Random House.

12 Managing volunteers

An application of ISO 31000:2009 risk management – principles and guidelines to the management of volunteers in tourism and beyond

Tracey J. Dickson

Introduction

Volunteers are essential for the effective operation of many businesses and activities across a community such as in sport, tourism and community organisations. People continue to volunteer despite concerns that they may have become time poor (Burgham and Downward, 2005; Edwards, 2005). The main reason for volunteering is knowing that their contribution is making a difference (Volunteering Australia, 2009a). Even though there are still people willing to volunteer, organisations continue to be concerned about their ability to recruit, train and retain volunteers (Volunteering Australia, 2009a). What may be needed is a more strategic approach to volunteer management that could have implications for the wider community through increases in social inclusion and social capital, as well as economic and cultural benefits that accrue from the volunteer contribution (Volunteering Australia, 2009b).

It is proposed that by ensuring that there is a sufficient supply of volunteers to meet the demand, and by appropriately managing the volunteers, there is a higher likelihood that they will return and/or continue volunteering in a variety of circumstances. This may help create a legacy for the organisation and the broader community into the future that will be of benefit to many. For example, it has been suggested that volunteering at a major sports event is more likely to raise interest in non-sports volunteering; with this in mind, investments in events are more likely to have wider social capital legacies (Downward and Ralston, 2006: 347). To achieve this legacy, appropriate volunteers first need to be recruited, then managed and encouraged to volunteer into the future. To this end, this chapter proposes the application of the international risk management standard ISO 31000 (Standards Australia and Standards New Zealand, 2009) as a framework for the management of volunteers in both the tourism context and beyond. ISO 31000 supersedes the third edition of the Australia and New Zealand Risk Management Standard (Standards Australia and Standards New Zealand, 2004).

Examples of the nexus between volunteers and tourism success may be seen from previous research that has investigated: guides in cultural institutions (Edwards, 2005); volunteers at mega sporting events (Bontempi, 2002; Fairley, Kellett and Green, 2007) and the provision of emergency services by volunteers to support tourist safety (Dickson, 2006; Uriely *et al.*, 2002). Examples of tourism events, attractions or activities that may depend upon the contribution of volunteers include:

- events, e.g. Jazz festivals, flower shows, motor shows
- cultural institutions, e.g. museums, art galleries
- sporting events, e.g. Olympics, World Masters Games, Sailing World Championships
- emergency services focusing on visitor safety, e.g. Surf Life Saving, Ski Patrol, Search and Rescue
- sport and recreation groups, e.g. Disabled WinterSport Australia in New South Wales, Australia uses around 100 volunteer guides each year to support people with disabilities to visit and enjoy the snow
- youth events, e.g. Scout Jamborees; World Youth Day.

The importance volunteers play in mega sporting is seen in that volunteers are central to many Olympic bids and essential to the Games' successful operation (Toohey and Veal, 2007). As indicated in Table 12.1, the average number of volunteers accredited by the Games' organising committee for the seven summer Olympics from 1980 to 2004 was 38,490, while for the eight winter Olympics from 1980 to 2006, the average number of volunteers was 14,544. For Beijing 2008, it was estimated that somewhere between 44,000 and 70,000 volunteers were recruited from almost 500,000 applicants (International Olympic Committee, 2008). For Vancouver 2010, 25,000 volunteers were recruited for the Winter Olympics and Paralympics (The Vancouver Organizing Committee for the 2010 Olympic and Paralympic Winter Games, 2009). These figures do not include other volunteers who may be helping on regional and cultural activities linked to the Games, but not run by the Game's organising committees.

Table 12.1 Volunteering at recent Olympics

Summer Olympics			Winter Olympics		
Year	Location	Volunteers	Year	Location	Volunteers
1980	Moscow	No data	1980	Lake Placid	6,703
1984	Los Angeles	28,747	1984	Sarajevo	10,450
1988	Seoul	27,221	1988	Calgary	9,498
1992	Barcelona	35,540	1992	Tignes	8,647
1996	Atlanta	47,466	1994	Lillehammer	9,054
2000	Sydney	46,967	1998	Nagano	32,000
2004	Athens	45,000	2002	Salt Lake	22,000
2008	Beijing (est)	44–70,000	2006	Torino	18,000
	Average of past games	38,490		Average of past games	14,544

Sources: IOC (2009) and Toohey and Veal (2007)

Should volunteers need to be replaced by paid employees the financial burden could be extreme and would require additional income sources to offset the costs. Estimates of the potential value of the volunteers' contribution includes:

- In 2005 the estimated value of the volunteer hours that ski patrollers contributed was $2,412 per volunteer or $159,186 for all respondents who were volunteers (Dickson, 2006);
- An estimate of the market value of volunteers at the 1999 World Ice Hockey Championships was 1.7 million euros for the 71,100 of volunteer hours (Solberg, 2003);
- The author estimates that if the 25,000 Winter Olympic and Paralympics volunteers in 2010 worked for 8–13 days each for 8 hours at a minimal $10 per hour, this would result in a wages bill in excess of C$20,000,000.

There has been extensive research into the motivations of volunteers across a wide range of settings (e.g. Dolnicar and Randle, 2007; Edwards, 2005; Farrell, Johnston and Twynam, 1998; Giannoulakis, Wang and Gray, 2008) and further research has explored the role of training and other management practices (e.g. Costa *et al.*, 2006; Cuskelly *et al.*, 2006; Farrell *et al.*, 1998). What this chapter adds to the discourse is a more strategic perspective to volunteer management that acknowledges volunteers as being essential to the achievement of the objectives of many tourism organisations and activities. There has been previous discussion to suggest that seeking to apply a commercially oriented human resources strategy to volunteers in on-profit sporting organisations may be excessive (Taylor and McGraw, 2006). However, given that there are many other tourism activities or attractions where volunteers are essential for the successful operation, and ultimately the commercial success, a more strategic approach to managing volunteers must be considered.

This chapter seeks to demonstrate how ISO 31000 (Standards Australia and Standards New Zealand, 2009) may be applied to the management of volunteers in tourism contexts in order to maximise the legacy to the person, the event, the attraction and the community. As there are many events, organisations and institutions that are dependent upon volunteers for their operation and success, the effective management of volunteers would be one aspect of their overall risk management. ISO 31000 is one of several recently released standards or standards in development (Standards Australia and Standards New Zealand, 1999, 2004; The Institute of Risk Management, 2002) that are aimed at a wide range of organisations, and that position risk as something that may have upsides as well as downsides. The first edition of the Australia and New Zealand Risk Management Standard (ANZRMS) was released in 1995, and was a world-first influencing those that followed culminating in ISO 31000.

ISO 31000

Defining risk

Previously the ANZRMS defined risk as 'the chance of something happening that will have an impact upon objectives' (Standards Australia and Standards New

Zealand, 2004: 4). ISO 31000 has simplified this to the 'effect of uncertainty on objectives' (Standards Australia and Standards New Zealand, 2009). This definition of risk differs greatly from many other definitions of risk that tend to focus on the potential for loss, yet as suggested previously, no one takes a risk in the expectation that it will fail (Bernstein, 1996). In fact, most will only take risks in the belief that there will be a return or a benefit that will arise. The definition used by the standard better reflects the dual nature of risk, where on one hand there is the sense of venturing or daring to take something on, which evolved from the 1557 verb *risicare*, versus the sense that there may be some danger, which evolved from the 1557 noun *risico* (Cline, 2003). The ISO 31000 definition encapsulates both the pursuit of desirable risks such as with financial investments, gambling, new ventures or adventure activities, but also the undesirable risks associated with disasters, crisis, environmental damage and event failures. It must be noted though that some risks are inherent to the activity and that to remove the risk may mean discontinuing the activity. For example, many outdoor activities have real physical risks, which may be minimised, but to truly eliminate them may mean stop doing the activity. Conversely, running an event has financial risks, but with effective management, the returns of taking those risks may be substantial.

For event and attraction managers in tourism who depend upon volunteers the risk may mean having insufficient volunteers for the tourism event/activity to proceed, or alternatively having the positive risk of an excess demand for volunteer positions, as has occurred with recent Olympic and Paralympic games.

Drawing upon the insights of triple bottom line reporting (e.g. Alpine Resorts Co-ordinating Council, 2009; Elkington, 1980; Global Reporting Initiative, 2006), the organisational objectives for tourism organisations, events or activities that may be impacted by uncertainty can cover a wide range of areas and across diverse time frames (Table 12.2). Volunteer management fits within and across these objectives.

Table 12.2 Foci of risk management within a triple bottom line framework

Economic	Social/community	Environment
Net profit	Staff turnover	Water usage
Business sustainability	Community perceptions	Power usage
Average visitor spend	Community engagement	Recycling
Visitor demand growth	Employment opportunities	Natural resource restoration
Year-round operations	for disadvantaged groups	Environmental degradation
	Visitor satisfaction or	Climate change policies
	experience	
	Marketing position	
	Corporate social	
	responsibility	
	Planning frameworks	
	Legal contexts	
	Image and reputation	
	Changing societal	
	expectations and norms	

Examples of the risks related to volunteers include both ensuring that there is sufficient demand for volunteers from within the organisation, as well as ensuring there is a sufficient supply of volunteers through recruitment, training, reward and recognition processes. Another area of risk is the legacy that is left afterwards, for example, for the London 2012 Olympics it has been suggested that after 2012,

> We aim to create a 'family' of volunteers after the Games who would like to stay in touch with friends made during the course of their volunteering. The Games will also have a far wider impact on the volunteering community in the UK.
>
> (The London Organising Committee of the Olympic Games and Paralympic Games Limited, 2009)

Equally, the legacy of volunteering for an event or at an attraction may be a decline in volunteering in the community more broadly if the volunteering experience was perceived as being negative or unsatisfying.

Defining risk management

ANZRMS defined risk management as 'the culture, processes and structures that are directed towards realising potential opportunities whilst managing adverse effects' (Standards Australia and Standards New Zealand, 2004: 4). This has been revised for ISO 31000 to 'coordinated activities to direct and control an organization with regard to risk' (Standards Australia and Standards New Zealand, 2009). This definition can include most things that are involved in running an organisation or activity from finances, to marketing, to strategic planning to community engagement. As noted in the introduction to the standard, 'risk management can be applied across an entire organization, to its many areas and levels, as well as to specific functions, projects and activities' (Standards Australia and Standards New Zealand, 2009). For volunteer management this may include a range of strategies such as having effective human resource strategies, internal marketing, reward and recognition systems and environmental scanning to monitor changes in the operating environment.

The following section explores the steps within the risk management process are explored, expanded and applied to the risk management of volunteers in tourism contexts.

The risk management process

ISO 31000 outlines a risk management process that may be adopted across a range of organisations, not just tourism organisations (Figure 12.1).

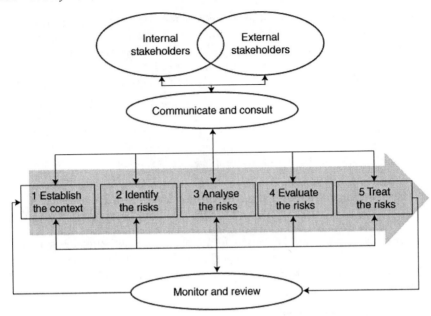

Figure 12.1 Risk management process adapted from AS/NZS ISO 31000:2009

Expanding the steps in the risk management process

1. Establish the context

> the organization articulates its objectives, defines the external and internal parameters to be taken into account when managing risk, and sets the scope and risk criteria for the remaining process.
>
> (Standards Australia and Standards New Zealand, 2009)

Understanding the context will influence how the risk management process is applied. Risk and risk management may differ in different social, legal, political and environmental contexts. Each tourism context is different, for example a mega-sporting event operates in a very different context and timeframe than a large national attraction, or a charity fundraising trek to the Himalayas or Kokoda, and each may also appeal to very different volunteers. Each activity may have different sets of criteria as to what risks may be acceptable, desirable or tolerable. An expedition to climb Mount Everest most probably has different levels of acceptable risk than an annual flower show, thus it is essential to be clear about what level of risks are acceptable to the organisation or activity with those people in that time and place.

In understanding the context it is also important to understand the motivations of both current and potential volunteers. Vining (1998) in a study of volunteers in schools identified a range of factors that 'turn-off' and 'turn-on' volunteers (Table 12.3). Understanding what motivates and demotivates volunteers in the diverse range of tourism contexts is essential for the effective management of their time and skills.

Table 12.3 Volunteers: turning them off and turning them on

Volunteer turn-offs	Volunteer turn-ons
Unclear responsibilities	Benefits to children
Poor organisation	Work that counts
Information withheld	Efficient organisation
Long meetings	Training and information
Tension and conflict	Friendly relations
Uncooperative staff	Efficient rosters
Burnout from overwork	Appreciation
	Recognition

Source: Vining, 1998: 43

2. Identify the risks

The organization should identify sources or risk, areas of impacts, events (including changes in circumstances) and their causes and their potential consequences.

(Standards Australia and Standards New Zealand, 2009)

Each organisation is unique, with a unique 'suite' of risks that reflects differences in their cultures, members, activities, objectives and timing. As indicated earlier, the objectives of an organisation may include economic, environmental and/or social objectives. As a result, what may impact upon these unique organisations with diverse objectives can also be equally varied. As an example of the range of things that may impact upon the supply of volunteers, during September 2009, the UK Home Office's Vetting and Barring Scheme, which aims to prevent anyone who is a risk to vulnerable groups from working with them, suggested a new strategy. It was suggested that the scheme should be expanded to include the need to screen parents who drive their children's friends to sporting events or social clubs. The organisations that depend upon these volunteers fear the policy will reduce the number of people volunteering and parents wishing to help their community (BBC News, 2009). By contrast, on the same day, it was reported that Michelle Obama, wife of the US President, promised to deliver a university commencement address if the students completed 100,000 hours of community service during the current academic year and thus increase the volunteer contribution in their community (Byers, 2009). As a further example, an economic downturn or financial crisis may have an impact upon the level of visitation to an event or attraction; however it may also increase the supply of skilled volunteers who have lost their paid employment. Conversely, a period of economic growth may increase the demand for the event or activity, but decrease the supply of volunteers as full employment levels are achieved across the economy.

Factors that may impact on an objective may also be the 'two sides of the same coin' (Figure 12.2), where a risk that is not managed well may have a negative impact; conversely if it is managed well it may have a positive impact upon the objective. In the example in Figure 12.2, the objective to grow an event that is dependent upon volunteer commitment will be negatively impacted if there are insufficient volunteers and equally may be positively impacted if there is a sufficient supply of appropriately skilled and experienced volunteers.

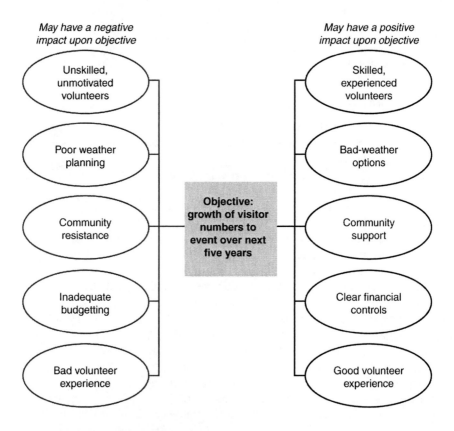

Figure 12.2 Examples of impacts upon an event objective

3. Analyse the risks

Analysing risk 'involves consideration of the causes and sources of risk, their positive and negative consequences, and the likelihood that those consequences can occur' (Standards Australia and Standards New Zealand, 2009). The analysis of the risks will help inform future decision-making, it is a combination of the source of the risk, the consequences should it occur (positive and negative) and the likelihood. The consequences and likelihood will be impacted by what existing controls are in place. Many events or activities will already have some existing controls in place to manage their risks, some may be formalised in documented policies and procedures, while others may be less formal and dependent upon word-of-mouth and prior experience.

In order to be able to effectively manage any risks, it is essential to understand the 'level' of risk, which is a result of the interrelationship between the likelihood that the event could occur, and the resultant consequences should it occur. Together this will give a risk score (Table 12.4) which then provides an indication of the level of intervention required given the context and risk criteria established earlier.

For an extreme risk, immediate preventative action may be required, for a high risk, senior management attention may be needed. For a moderate risk, middle management responsibility may be needed, while for low risks management may be by routine procedures. If, however, an organisation is micro or small, all these responsibilities may be collapsed down to being one aspect of an individual or the owner's overall role in the organisation.

In the following tables, the risk analysis matrix is applied to the supply of (Table 12.5) and demand for (Table 12.6) volunteers. In these two examples, the matrix has been expanded to demonstrate how it is possible to include the risk management strategies, the remaining or residual risk score following the application of the risk management strategies, and who within the organisation may be responsible for the actions.

Table 12.4 Risk analysis matrix

	Consequences				
Likelihood:	*Insignificant 1*	*Minor 2*	*Moderate 3*	*Major 4*	*Catastrophic 5*
A (Almost certain)	A1 High	A2 High	A3 Extreme	A4 Extreme	A5 Extreme
B (Likely)	B1 Moderate	B2 High	B3 High	B4 Extreme	B5 Extreme
C (Possible)	C1 Low	C2 Moderate	C3 High	C4 Extreme	C5 Extreme
D (Unlikely)	D1 Low	D2 Low	D3 Moderate	D4 High	D5 Extreme
E (Rare)	E1 Low	E2 Low	E3 Moderate	E4 High	E5 High

Table 12.5 Example of the risk analysis matrix as applied to the supply of volunteers

Risk	Likelihood	Consequence	'Risk score'	Actions required	Residual risk	By whom?
1 Inadequate or inaccurate marketing of volunteer opportunities	C Possible	3 Moderate	C3 High	Develop and monitor a Volunteer Strategic Plan in conjunction with HRM	C2 Moderate	HR
2 Increased competition for the 'volunteer hour' from other activities	B Likely	3 Moderate	B3 High	Ongoing celebration and recognition of volunteers via internal and external means	B2 High	Supervisors or line managers
3 Public transport to work sites reduced restricting volunteers' access	C Possible	2 Minor	C3 High	Monitor public transport plans Communicate with local transport providers and council	C2 Moderate	HR
4 Declining societal support for volunteerism	D Unlikely	4 Major	D4 High	Ongoing publicity about the contribution made by Vols in media, internal publications and reports	C2 Moderate	Marketing or PR
5 Failure to provide safe work systems for volunteers	C Possible	3 Moderate	C3 High	Train Vol Co-ord in OHS roles and responsibilities Train Vols in OHS roles and responsibilities Implement Vol Mentors to assist in support of Vols	C2 Moderate	HR
6 Volunteers' skills and experiences undervalued and/or under-utilised	B Likely	2 Minor	B2 High	Skills analysis of all Vols linked to Strategic Plan and entered onto database Implement Vol Mentors to assist in support of Vols	D2 Low	HR Line managers

Table 12.6 Risk analysis matrix applied to the internal demand for volunteers

Risk	Likelihood	Consequence	'Risk score'	Actions required	Residual risk	By whom?
1 Volunteers not seen as a strategic part of organisation's work	D Unlikely	3 Moderate	D3 Moderate	Ensure Vols are recognised in all strategies and reports	D2 Low	CEO / Board
2 Legislation (e.g. Child Protection, OHS) making it prohibitive to use volunteers	C Possible	4 Major	C4 Extreme	Monitor legislative climate Lobby as necessary	C2 Moderate	HR and CEO
3 Insurance premiums for volunteers becoming excessive	C Possible	4 Major	C4 Extreme	HR to monitor insurance climate Lobby as necessary	C2 Moderate	HR / Finance
4 Paid staff unsure of how to maximise the use of volunteers	C Possible	3 Moderate	C3 High	In-service staff on volunteer opportunities Increase marketing of the role of volunteers Promote the volunteers' contribution via internal and external marketing strategies	C2 Moderate	HR / Marketing / PR

4. Evaluate the risks

> The purpose of risk evaluation is to assist in making decisions, based on the outcomes of risk analysis, about which risks need treatment and the priority for treatment implementation.
>
> (Standards Australia and Standards New Zealand, 2009)

Depending upon the risk criteria established in Step 1, the risk matrix will help clarify those areas of risk that are acceptable (such as low or moderate risks) or unacceptable (such as high and extreme) for the organisation, event or activity within a particular time and place (Table 12.7). For volunteers, it may be considered acceptable to the organisation to have a few less volunteers than planned, but not acceptable for volunteers to be underutilised in their roles or not recognised or rewarded for their contribution.

5. Treat the risks

> Risk treatment involves selecting one or more options for modifying risks, and implementing those options.
>
> (Standards Australia and Standards New Zealand, 2009)

The risk management strategies may impact upon the likelihood, the consequence, or both, of the identified risks, and thus the resultant risk score (e.g. Table 12.5, Table 12.6 and Table 12.7), but as noted in ISO 31000, the strategies need to be cost-effective. To do otherwise may result in unacceptable financial risks.

Table 12.7 Risk analysis matrix: examples of what may be deemed to be acceptable levels of risk

	Consequences				
Likelihood:	*Insignificant* 1	*Minor* 2	*Moderate* 3	*Major* 4	*Catastrophic* 5
A (Almost certain)					
B (Likely)			NOT ACCEPTABLE		
C (Possible)					
D (Unlikely)	ACCEPTABLE				
E (Rare)					

Communicate and consult, and monitor and review

> Communication and consultation with internal and external stakeholders should take place during all stages of the risk management process

and

> Both monitoring and review should be a planned part of the risk management process and involve regular checking or surveillance.
>
> (Standards Australia and Standards New Zealand, 2009)

The final two phases discussed in ISO 31000 are essential for continuous improvement and to ensure that the risk management process reflects and adapts to changing circumstances, internal and external. Communication will assist with developing effective risk management strategies that have the support of key stakeholders as well as gaining important buy-in from stakeholders, while the monitoring and reviewing will aid learning from the process and contributing to the continuous improvement of the developing risk management process and to identify and respond to the changing context and risk culture.

Examples and questions that may be considered in the risk management process in relation to volunteers are summarised in Table 12.8 (overleaf). The questions and examples have been outlined for each of the steps (1–5) in order to guide the process.

Conclusion

Volunteers are an essential resource for many tourism events, activities or organisations (Figure 12.3). Their contribution adds to the social fabric of a society and the effective management of volunteers may enhance the contribution they make across a community as a whole. The appropriate management of volunteers may benefit from the application of the *Risk Management Process*. Risk management is not about avoiding loss, but it *is* about managing for uncertainty in organisations that may have 'upsides' as well as 'downsides'. The risk management process acknowledges that risks may have both positive and negative impacts upon an organisation's objectives and that risk management aims to minimise the negative impacts and maximise the positive impacts. By understanding the context in which the organisation or activity is operating, establishing what level of risks are acceptable, identifying and analysing the risks, then clear treatments may be established that are relevant to the unique situation. The insights gleaned from the examples presented here highlight how risk management of volunteers is a whole-of-organisation activity and may apply across all levels of an organisation or activity.

Table 12.8 Steps in the risk management process

Step	Questions and examples
1 Establish the context	What is the previous volunteering experience of the organisation/staff? What is the culture of volunteering in the broader community? Who is likely to be a volunteer: demographics, interests, skills, qualifications, previous paid and volunteer experience? What other demands (regular and irregular) are there for the volunteers' time from other groups in the community, sporting, charitable arenas? What encourages/discourages people to volunteer? What do the volunteers want from the experience? Who ISN'T volunteering and why?
2 Identify the risks	Insufficient supply of volunteers from the target community Insufficient demand for volunteers from within the organisation or event Inappropriate management of volunteers impacting upon their willingness to volunteer again Bad volunteering experiences elsewhere impacting upon volunteers' willingness to volunteer for you Legislative changes such as with Child Protection or Occupational, Health and Safety that makes using employees more difficult Public transport changes that make it difficult for volunteers to travel to the event or activity Government policies that promote the role of volunteers and increase the supply of volunteers
3 Analyse the risks	In relation to the supply of volunteers, questions that could be considered include: • How do you ensure that there are sufficient appropriately skilled volunteers available when the work demands? • How likely is it that there will be *insufficient* volunteers available? • Would the event / organisation survive without volunteers? Questions in relation to the demand for volunteers include: • What internal marketing is there to promote the contribution of volunteers and hence the internal demand for volunteers? • Is there sufficient meaningful and challenging work available for volunteers to draw upon their skills and qualifications as well as development opportunities?

continued

4 Evaluate the risks	Drawing upon Vining (1998), it is also important to consider what may happen if,
	There is an excess of volunteers for the work available, how do you encourage the individual to be motivated to offer to volunteer again in the future?
	There were insufficient volunteers, what it would cost to recruit, train and pay for qualified staff?, e.g.
	• Volunteers do not feel valued, or that they are not making a meaningful contribution and therefore leave early or do not return in the future?
	• Volunteers don't want to volunteer in the future (for you or others) as a result of a unsatisfying volunteering experience?
	• What is the legacy of volunteering that you are leaving for other organisations, activities or the broader community?
	• Paid staff do not have the skills to manage a flexible, volunteer workforce?
5 Treat the risks	Internal marketing of the benefits that volunteers bring, and to identify volunteer opportunities
	Analysis of the skills of volunteers to ensure appropriate task allocation
	Training of permanent staff to work with volunteers to maximise the opportunities for both
	Job descriptions for volunteers to ensure clarity of expectations
	Rewards and remuneration for volunteers, e.g.
	• Badges for service recognition;
	• Celebration events for volunteer contribution;
	• Articles in newsletters recognising the role of volunteers;
	• Networking with like organisations or activities to share volunteer management strategies.

Figure 12.3 It's a long way to the top without your volunteers! Former Australian
Paralympians ascended Mt Kosciuszko on the International Day for People
with Disabilities, December 2008, with the support of a team of volunteers

References

Alpine Resorts Co-ordinating Council (2009) *Social Key Performance Indicators for Victoria's Alpine Resorts*. Melbourne: Alpine Resorts Co-ordinating Council.

BBC News (2009) NSPCC criticises volunteer checks, *BBC News*. Available from http://news.bbc.co.uk/2/hi/uk_news/8253099.stm. Accessed 15 September 2009.

Bernstein, P. L. (1996) *Against the Gods: The Remarkable Story of Risk*. New York, NY: John Wiley and Sons.

Bontempi, R. (2002) The ideal Olympic volunteers. *Olympic Review*, 23–26. Available from http://www.la84foundation.org/OlympicInformationCenter/OlympicReview/2001/OREXXVII42/OREXXVII42o.pdf. Accessed 6 January 2009.

Burgham, M. and Downward, P. (2005) Why volunteer, time to volunteer? A case study from swimming. *Managing Leisure*, 10(2): 79–93.

Byers, A. (2009) *First Lady Promises Commencement Address if George Washington University Students Volunteer*. Available from http://www.foxnews.com/politics/2009/09/11/lady-promises-commencement-address-george-washington-university-students. Accessed 15 September 2009.

Cline, P. B. (2003) *Re-examining the Risk Paradox*. Paper presented at the Proceedings of the 2003 Wilderness Risk Management Conference, State College, PA.

Costa, C. A., Chalip, L., Green, B. C. and Simes, C. (2006) Reconsidering the role of training in event volunteers' satisfaction. *Sport Management Review*, 9(2): 165–182.

Cuskelly, G., Taylor, T., Hoye, R. and Darcy, S. (2006) Volunteer management practices and volunteer retention: a human resource management approach. *Sport Management Review*, 9(2): 141–163.

Dickson, T. J. (2006) *Patroller Survey – 2006, A Report Prepared for the Australian Ski Patrol Association*. Canberra, ACT: Centre for Tourism Research, University of Canberra.

Dolnicar, S. and Randle, M. (2007) What motivates which volunteers? Psychographic heterogeneity among volunteers in Australia. *Voluntas: International Journal of Voluntary and Nonprofit Organizations*, 18(2): 135–155.

Downward, P. M. and Ralston, R. (2006) The sports development potential of sports event volunteering: insights from the XVII Manchester Commonwealth Games. *European Sport Management Quarterly*, 6(4): 333–351.

Edwards, D. (2005) It's mostly about me: reasons why volunteers contribute their time to museums and art museums. *Tourism Review International*, 9: 21–31.

Elkington, J. (1980) *The Ecology of Tomorrow's World*. New York: Halsted Press.

Fairley, S., Kellett, P. and Green, B. C. (2007) Volunteering abroad: motives for travel to volunteer at the Athens Olympic Games. *Journal of Sport Management*, 21(1): 41–57.

Farrell, J. M., Johnston, M. E. and Twynam, G. D. (1998) Volunteer motivation, satisfaction, and management at an elite sporting competition. *Journal of Sport Management*, 12(4): 288–300.

Giannoulakis, C., Wang, C. H. and Gray, D. (2008) Measuring volunteer motivation in mega-sporting events. *Event Management*, 11: 191–200.

Global Reporting Initiative (2006) *Sustainability Reporting Guidelines v. 3.0*. Available from http://www.globalreporting.org/NR/rdonlyres/ED9E9B36–AB54–4DE1–BFF2–5F735235CA44/0/G3_GuidelinesENU.pdf. Accessed 29 July 2009.

International Olympic Committee (2008) Nice to meet you, Zhang Jiayu. *News about Beijing 2008*. Available from http://www.olympic.org/en/content/Media/?FromMont h=August&FromYear=2008&ToMonth=August&ToYear=2008&currentArticlesPag eIPP=10&currentArticlesPage=8&articleNewsGroup=-1&articleId=53440. Accessed 26 August 2009.

International Olympic Committee (2009) Official Website of the Olympic Movement, from http://www.olympic.org. Accessed 29 July 2009.

Solberg, H. A. (2003) Major sporting events: assessing the value of volunteers' work. *Managing Leisure*, 8(1): 17–27.

Standards Australia and Standards New Zealand (1999) *AS/NZS 4360: 1999 Risk Management*. Strathfield, NSW: Standards Association of Australia.

Standards Australia and Standards New Zealand (2004) *AS/NZS 4360: 2004 Australian/ New Zealand Standard: Risk Management*. Strathfield, NSW: Standards Association of Australia.

Standards Australia and Standards New Zealand (2009) *ISO 31000 Risk management – Principles and guidelines*. Sydney, NSW: Standards Australia and Standards New Zealand.

Taylor, T. and McGraw, P. (2006) Exploring human resource management practices in nonprofit sport organisations. *Sport Management Review*, 9(3): 229–251.

The Institute of Risk Management (2002) A Risk Management Standard. Available from http://www.theirm.org/publications/PUstandard.html. Accessed 4 November 2009.

The London Organising Committee of the Olympic Games and Paralympic Games Limited (2009) The Volunteer Programme. *The Official Site of the London 2012 Olympic and Paralympic Games*. Available from http://www.london2012.com/get–involved/ volunteering/the–volunteer–programme.php. Accessed 20 January 2009.

The Vancouver Organizing Committee for the 2010 Olympic and Paralympic Winter Games (2009) *Volunteer Opportunities*. Available from from http://www.vancouver2010. com/en/work–and–volunteer/volunteer–opportunities/–/33922/w6niz7/index.html. Accessed 23 January 2009.

Toohey, K. M. and Veal, A. J. (2007). *The Olympic Games: A Social Science Perspective* (2nd edn). Wallingford, UK: CAB International.

Uriely, N., Schwartz, Z., Cohen, E. and Reichel, A. (2002) Rescuing hikers in Israel's deserts: community altruism or an extension of adventure tourism? *Journal of Leisure Research*, 34(1): 25–37.

Vining, L. (1998) *Working with Volunteers in Schools: Understand Volunteers and Reap the Benefits for Your School*. Carlingford, NSW: Lenross Publications.

Volunteering Australia (2009a) *National Survey of Volunteering Issues 09*. Available from http://www.volunteeringaustralia.org/files/CFBMHJNYX5/FINAL_National_Survey_ of_Volunteering_Issues.pdf. Accessed 16 December 2009.

Volunteering Australia (2009b) Volunteering Research Framework. Available from http:// www.volunteeringaustralia.org/files/V4J4N0BTTA/Research%20Framework%20 Final%20for%20Web.pdf. Accessed 16 December 2009.

13 Volunteer tourism and intercultural exchange

Exploring the 'Other' in the experience

Stephen Wearing and Simone Grabowski

Volunteer tourism as a form of international development has been posed as an alternative mechanism which has the potential to achieve different socio-cultural outcomes. In this guise it aims 'to establish direct personal/cultural intercommunication and understanding between host and guest' (Dernoi, 1988: 89). This chapter explores the volunteer tourist and their interaction with the host community. It is argued that the relationship between the volunteer tourist and the community gives shape to a richer understanding of the volunteer tourism experience, where more equal power relationships are evolving and where the experience is more inclusive of the 'Other'. Where tourism in less developed countries is frequently criticised as creating development that results in power inequalities between host and guest it is important to examine this issue for volunteer tourism. The cultural exchange with those who are 'othered' by the mainstream tourism experience is the basis for a discussion that highlights the complexity of the relationship between hosts and guests. Within the limited literature on volunteer tourism, it is suggested that these tourists have very different motivations for travel compared with the more traditional tourists or mass tourists. This chapter will begin by providing background on volunteer tourism and follow on with presenting a theoretical understanding of the 'other' and cultural exchange under an inclusive research paradigm in the context of tourism. This will be supported by a case study of Youth Challenge Australia volunteers. Finally, a discussion is offered with reference to the alternative mechanisms that are developed to engage youth in volunteering for development.

Introduction

The culture of a host community is often commoditised by the tourism industry or discarded by the tourist as a genuine motivation for travel. In a debate that has spanned over thirty years, the tourist has either been seen to seek authenticity or meaning in the real life of others in the host community (MacCannell, 1976; 1999; Rojek, 1997; Urry, 1990) or a shallow experience unlikely to leave lasting impressions (Turner and Ash, 1976). This debate has been refuelled in the last few years by academics interested in a growing niche market: volunteer tourism.

Volunteer tourism is estimated to attract 1.6 million people worldwide annually (TRAM, 2008). Broadly defined it is travel which involves 'aiding or alleviating the material poverty of some groups in society, the restoration of certain environments or research into aspects of society or environment' (Wearing, 2001: 1). It is usually undertaken by Western youth between the ages of 18–30 in developing countries where the culture is vastly different from that to which they are accustomed.

Volunteer tourists have been found to display several altruistic and self-serving motivations. Brown and Lehto (2005) found that the three principal motivations behind volunteering were cultural immersion, making a difference and seeking camaraderie. Broad and Jenkins (2008) found that the motivations overlap between an altruistic desire to help wildlife, a need for travel, career development and personal development. These motives are very different from those explored in mainstream tourism literature and voluntary action literature. For example, Pearce (2005) identified the core travel needs as rest/relaxation, explore and social interaction, while Martinez and McMullin found six variables that are associated with the volunteer's desire to fulfil 'higher-level' needs. These are 'efficacy, personal motivation, request, social networks, lifestyle changes, and competing commitments' (2004: 116).

The fact that the activity is often performed in developing countries by Western youth means that there will be inherent cross-cultural issues in the tourist experience. These issues and the experiences the volunteer tourists have in the host destination have only just begun to be examined in the tourism literature (Laythorpe, 2009; McIntosh and Zahra, 2007). Although initially the interaction was said to have a very positive effect on the communities, this is now being questioned by those who see the experience to be of greater benefit to the volunteer (Ehrichs, 2000; Ingram, 2009; Raymond and Hall, 2008; Simpson, 2004; Sin, 2009). This chapter will discuss some of these issues and present the notion that the characteristics of the volunteer tourist as a market segment minimise the amount of 'othering' that traditional tourism has been criticised for.

Development of a decommodified research agenda

Wearing, McDonald and Ponting (2005) argue that Western, neoliberal, free market paradigms continue to dominate the tourism research agenda. They contend that alternative research paradigms are needed to enrich the field and to provide new ways of seeing, researching and doing tourism. Decommodified research paradigms, based upon feminist theory, ecocentrism, community development and post-structuralism, offer more depth to enable examination of this area. A decommodified approach to tourism research opens the way for exploration of tourism's potential to provide the means for community-defined and community-driven development and conservation. This chapter examines how intercultural exchange can occur in volunteer tourism based around these central tenets providing mechanisms that move away from the commodified free market regime that has existed. In calling for a widening of the paradigmatic scope, models of tourist–host contact can be developed for volunteering that are more equitable than those underpinned by the view of a simple economically rational exploitation of an experiential commodity.

Some authors (cf. Butcher, 2005; 2006) argue that alternative types of tourism like volunteer tourism are or will become mainstream. They are criticised as being overly moralistic in their approaches and consequently are discredited as being able to provide sustainable alternative experiences. However, through organisations such as non-government organisations (NGOs) the theorisation of the discourse and practice of volunteering and sustainable tourism can provide a rich ground for understanding the cross-cultural exchange between host and guest. It can ascribe new meanings and practices particularly relevant in the Global Financial Crisis (GFC) of 2008/9 where an alternative mechanism of understanding from market economy centralisation is essential in the move to a more sustainable future. This is not an approach, as Butcher (2006) contends, that can be renounced as an economically untenable replacement of 'the market' with charity but it is a way of incorporating the host community into research where they are moved beyond the 'other' in the tourism experience. Butcher's arguments remain within the mainstream tradition of tourism research, a tradition that sits within a commodified framework. He has voiced opposition to a mistaken assumption that advocacy of alternative research paradigms is anti-development, and he proclaimed his position that 'the market' (Western, free market, neoclassical economics) is the only paradigm of tourism research that makes sense: that there is no way of valuing quality of life, nature, standards of living or development other than by the conventional economic indicators that are of Western design. In this view host communities need help to 'liberate themselves from their environment' because their close links with the natural environment characterise their impoverishment (Butcher, 2005: 122). Though presented by Butcher as a singular truth, this view has never been articulated by the rural indigenous communities that the authors have worked with in Australia, Papua New Guinea, Indonesia, Costa Rica or Fiji. Indeed, the Western ethnocentrism at its core has been widely criticised for failing to consider the wants, ambitions and priorities of those about to be subjected to this process of 'Westernisation' (Brohman, 1996; Galli, 1992; Harrison, 1992a; Mehmet, 1995; Said, 1978; 1993; Schmidt, 1989; Telfer, 2002; 2003; Wiarda, 1988). Ironically, the academically stifling nature of that framework illustrates why there is a need for more radical alternatives in the research agenda for tourism. In addressing the volunteer cross-cultural experience with the host community the researchers suggest that there needs to be a move beyond the current paradigm.

Despite claims to the contrary, a decommodified tourism research agenda does not advocate withdrawal from global tourism markets and networks; clearly no tourism venture can survive in isolation from the rest of the industry. It does, however, enable exploration of the notion that where something like volunteer tourism is concerned people are interested in purchasing a tourism product (through the normal distribution channels of the broader tourism industry) that is not commodified to the point where the experience has become homogenised and where the concerns of destination communities have been left out of the planning, operationalisation, marketing and distribution of tourism's benefits (cf. Ponting, McDonald and Wearing, 2005).

The decommodified approach advocated here aims to explore development through tourism as something that is defined by local communities in their own terms and which may then allow the exploration of the 'othered' in this experience in a more intricate manner. The opportunities offered by this approach allow a more thorough engagement and analysis that can seek to increase understanding of communities in developing countries and the intercultural exchanges that occur between the volunteer and the community. If this approach is neglected the opportunity to increase understanding of tourism in these communities is missed. Understanding local ways of valuing enables a more sophisticated discussion than those offered by the economic rationalism of developed counties. Postcolonial criticism, for example, enables us to destabilise and deconstruct Eurocentric, homogenising notions of the coloniser and the colonised. In Eurocentric theory, the coloniser was assumed to be superior, and the coloniser's knowledge and formulation of history was assumed to be universal, correct and rightfully dominant. Postcolonial criticism has usefully forced 'a radical rethinking of forms of knowledge and social identities authored and authorised by colonialism and Western domination' (Prakash, 1994: 87). Similarly, we advocate that the decommodified research agenda encourages a radical rethinking of the dominant, Western, neoliberal economic approach to tourism research. A view which sees development as urbanisation and modernisation, ecocentrism as a threat to a modernist (humanist) worldview, and which prematurely dismisses alternative approaches to tourism research is symptomatic of the mainstream tradition of tourism research. This chapter then will pursue the inclusion of the 'othered' into the understanding of intercultural experiences by volunteer tourists.

Power, knowledge and the other

Attempting to decommodify the traditional approach to tourism is difficult. Jack and Phipps argue that the Western tourists 'seek solace in the Other, where the Other is viewed as an equally homogeneous, but diametrically opposed set of people that are untainted by the violent orderings of modernity' (2005: 24). Foucault's (1980; 1986) illustration of the historical process of 'othering' documented how 'out-groups' were pushed to the periphery of society, both physically and metaphorically. These ideas have been applied to tourism studies (cf. Cheong and Miller, 2000; Wearing and McDonald, 2002). Such studies illustrate that the process of commodification inherent in the development of tourism leads to the segregation and exclusion of local communities from participating in or sharing the process, functions and economic benefits of the industry. Tourism here is shown to be a 'punitive', 'disciplinary' exercise where space is controlled through the articulation and combination of forces. These include the demarcation of tourist space and the creation of a periphery (typically used to house workers and to act as a transportation hub to bring tourists and goods into the tourist site), through financial investment tied to specific commodified outcomes, the application of

Western models of professional management, power/knowledge and language, and the employment of Western tourism operators (Ponting *et al.*, 2005). In these situations local communities are theoretically, physically and economically excluded by the tourism industry. This then creates the basis for these communities to be 'othered' in the tourism experience and raises the issues that surround the ideas inherent to volunteer tourism and intercultural exchange. If the volunteer is to become a part of the host community's space they must move beyond the traditional demarcations that do not acknowledge how these communities are 'othered' under traditional tourism experiences and research. How then is it possible to understand and explain what occurs in the volunteer tourist's experiences?

Often destination images do not match the tourist experience because the voices of the local people have long been silenced. The large majority of the images portrayed in tourism marketing materials are constructed without the participation of the local communities who in turn have potential to confer social value (Cunningham, 2006; Macleod, 2006). Social valuing recognises that local communities hold extensive knowledge about areas and an exposure to this knowledge can play a key role in the tourist experience provided of course that the locals are in control of the interpretation and transmission of this knowledge. The successful communication of social valuing can enable the tourist to transcend the Otherness implied and represented in many tourism marketing images. The involvement of local communities in the marketing of their cultures allows for a greater range and diversity of images, messages and symbols to be communicated. In this alternative process, the local community becomes pivotal to the success of the tourism venture. Social value thus becomes key to the marketing of the tourist destination with locals playing a central role in determining the 'identity' of their destination through the value they ascribe to particular places, events and traditions.

Tourism is often unplanned and disorganised. This can be, and in some cases has been, rectified by the involvement of all levels of government, NGOs and locals working together to achieve sustainable outcomes that focus on the wishes of the community. If tourism is produced and distributed at the local level by local communities it has a much greater chance of resisting the global imperatives of capital intensification (MacCannell, 2001; Ritzer, 2006; Sofield, 2003; Yudice, 1995). Indeed in developing countries the involvement of government planners in tourism is seen as crucial for building and mediating local social value (Wilson, 1997). Sofield (2003) argues that state participation is vital in the tourism planning process in order to manage the balance of power between central control and local empowerment, particularly when external developers and investors are involved. Sofield (2003) cites the example of the collapse of a tourism resort on Anuha in the Solomon Islands where the central government failed to act in a dispute between the local community and the new owners of an existing resort who wanted to expand without local community consultation.

Haywood (1988) argues that if tourism is not planned and organised in a way that is sensitive to the local community then tolerance thresholds of that community are likely to be exceeded. When a breakdown in the relationship between local communities, tourism brokers and tourists occurs the result is often conflict and the tourism industry may 'peak, fade and self destruct' (Haywood, 1988: 105). Conflicts between locals and tourists typically result from capital intensification and economic development of the destination which is viewed as a vehicle that will disrupt and dilute local culture (Robinson and Boniface, 1999). Conflict is also likely to arise over the use of scarce resources for tourism, when locals are denied access to the natural resources upon which they base their livelihoods. Tourists may also 'display ignorance ... or ... disregard for the environment and indulge in inappropriate [and culturally insensitive] behaviour' that angers (and disempowers) the local community (Holden, 2000: 74).

For example, the onset of large-scale tourism (or conventional mass tourism CMT) in Goa on the western coast of India has produced pressures on both society and the environment (Brammer, Beech and Burns, 2004). Reactions to this increase in tourism have been varied, but organised forms of stakeholder resistance have become common. Major issues that have emerged include the community's reaction to disputes over the use of land and, in particular, the use (and abuse) of beaches by tourists and tourism operators. Conflict in Goa is centred on the main stakeholders – the small-scale entrepreneurs who seek to make a living from tourism through running beach shacks, hawking goods and organising rave parties, and the large corporate interests whose developments include beach-front hotels and casinos, and who see the market as an unsophisticated extension of 'sun lust' tourism by Europeans. Brammer *et al.* (2004) argue that these and other conflicts stem largely from a lack of adequate planning, consultation and mediation between the various stakeholders by the Goan authorities – failure to value the place and its cultures.

Conventional mass tourism (CMT) has been criticised for damaging society through the commodification of culture (Harrison, 1992a; Mathieson and Wall, 1982). The culture of the destination is exposed to tourists through the display of religious and tribal rituals or the selling of traditional arts and crafts through an increased interaction between locals and tourists. MacCannell (1973) has proposed that these cultural displays have the potential to become staged and lose their meaning for the host populations. Additionally, Valentine (1992) suggests that local communities begin to resent tourists who in many cases are more affluent than the local people. They have different religious and cultural backgrounds and portray a lack of respect for the local culture wearing offensive clothing or entering restricted religious sites. Young people in local communities begin to follow these displays which is noted as the 'demonstration effect' (Harrison, 1992b; Macleod, 2004; Teo, 1994). This results in greater social problems such as crime, drugs and prostitution (Holden, 2000).

Some recent empirical studies have argued against a theory of cultural homogenisation and subjugation of the host community. Macleod (2004), for example, concludes that the influx of many different cultural groups to an isolated community can increase awareness of the diversity and the host community can 'become increasingly aware of their individuality and group identity' (2004: 218). Additionally, Doron's (2005) study of the boatmen on the Varanasi has shown that although unequal power relationships do exist between the tourists and the boatmen, it is unclear who is the 'subordinate' player. He concludes that in the exchange that takes place, the tourists are in fact 'othered' by the boatmen. These are just two examples of a community's response to tourism where they are able to negate or reverse the traditional power subjugations that occur when two cultures collide. As an exchange also takes place in volunteer tourism, it is argued that the outcome and relationship between volunteer and community is not one of hostility but one of mutual benefit.

Cultural exchange and adaptation

Culture is a very difficult concept to define. Traditionally it was a term to denote Western civilisation and in more modern times has been substituted for popular music, art and literature or 'high culture' (Hall, 1997). However, in a socio-anthropological sense, culture has been defined as the shared values of a group or of a society that guide behaviour (Hall, 1997; Jandt, 2007). Bakhtin, Emerson and Holquist (1986) describe how a culture does not have any meaning until it comes into contact with another. The meeting of these two cultures does not result in merging or mixing; instead, each is mutually enriched. In a world where international borders have been all but removed, political, social, economic and technological contact occurs on a daily basis between different cultures. Some of the groups that are exposed to cross-cultural contact are international students, business travellers, diplomats, military personnel, migrants, refugees and tourists. They encounter different cultures out of necessity, because of opportunity and increasingly due to ease of movement. Bochner (1982) describes the contact as creating an 'us' versus 'them' platform where both groups see a difference in physical appearance, language and religion in the other. This has caused conflict throughout the world; however, it has also fostered learning and exchange.

Theorising intercultural exchange has been the focus of anthropologists, sociologist and socio-psychologists for over thirty years. In this time several models have been produced to explain the cross-cultural experiences and adaptation processes of sojourners. Intercultural adaptation is the process whereby people adapt their behaviour to facilitate understanding in cross-cultural situations (Reisinger, 2009). Ellingsworth theorised that 'intercultural communication is viewed as occurring under conditions often characterized by disparity of purpose, inequality in status and power, and advantage related to setting' (1983: 203). His Intercultural Adaptation Theory (IAT) proposed that adaptive behaviour will be shared only when equity is present, otherwise the burden of adaptation shifts towards the less advantaged. Thus true intercultural communication can only occur between groups that strive to achieve the same goals and the power relations between them are equal (see Figure 13.1). Cai and Rodriguez (1997) build on this theory to produce an Intercultural Adaptation Model (IAM). They propose that misunderstanding occurs when there are tenuous preconceived notions about the culture or there is a lack of shared knowledge between participants in the exchange process. This makes the adaptation process very difficult. If the outsider is prepared then they can avoid cross-cultural misunderstandings and more importantly, effective adaptation and increased understanding will occur as a result of the positive experiences. They argue that the experience plays a central role in the process of cross-cultural adaptation.

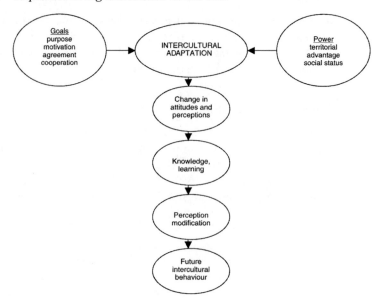

Figure 13.1 Intercultural Adaptation Theory (IAT)

Source: adapted from Reisinger (2009: 61)

Ward, Bochner and Furnam (2001) note that culture shock is another result of intercultural communication. Furnham (1984) discusses that culture shock is a consequence of the language difficulties experienced by the sojourner or even the cross-cultural differences in social skills. These differences can cause great social difficulty for the sojourner and can be simple, everyday activities like ordering food, shopping or making contact with the opposite sex (Furnham and Bochner, 1982). Research into the effects of culture shock and intercultural exchange is limited in tourism literature with Weaver (1987) claiming that 'tourists seldom experience culture shock because they are short-term sojourners who never actually enter another culture'; however, this is challenged by Pearce (1981, cited in Furnham and Bochner, 1986) whose study of Australian holiday-makers resulted in a suggestion that the environmental change ('shock') may indeed have negative consequences on the tourists. From a community point of view, this exchange has been explored in early anthropological work on host-guest relations in tourism where, in small numbers, the tourists are viewed as anthropologists (cf. Smith, 1989). Ward *et al.* (2001) explain that it is in the niche markets like ecotourism and backpacker tourism that the cultural adaptation of tourists and the impact they have on communities can be more thoroughly explored. The reason for this is that these tourists have greater contact and spend longer periods of time with the host communities. The last twenty years has seen the growth of these new types of tourism which are small in scale, independent and self-sustaining – entirely the opposite to the mass packaged tours made popular in the twentieth century. Hunter and Green note that 'tourists are becoming more discerning, seeking activities, arrangements and experiences which depend, crucially, on a high-quality physical and cultural environment' (1995: 7). Additionally, Sofield (1991) points out that the prospect of encountering different cultures attracts tourists to different destinations. This tourism has been given many names: responsible tourism (Wheeller, 1991), ecotourism (Wearing and Neil, 1999), new tourism (Mowforth and Munt, 2003) and alternative tourism (Mieczkowski, 1995) to name a few. They all share a common interest in ensuring minimal impact and achieving 'sustainability' and they also rely on a deeper exchange between host and guest.

It has been argued that these alternative forms of tourism are in fact just as harmful as the CMT that preceded them purely due to the close contact between host and guest (Cater, 1993; Macleod, 2004). We recognise that the values and existing power relationships found with any form of tourism to developing countries will always reproduce social inequities. Therefore, there is a need also to examine if volunteer tourism only serves to perpetuate, or even exacerbate, racial and ethnic stereotypes. In order to fully appreciate the exchange that occurs in this experience it is suggested that the learning experience is a vital outcome and comes about in the membership and participation in community practices (Lave and Wenger, 1991). In order to understand the volunteer experience it is essential to appreciate how the intercultural exchange in this experience works, which appears to be difficult using current models found of the tourism experience. According to Furnham's (1984) 'Linear Relationship between Length of Stay and Degree of Stress', the socio-psychological effects of the cultural experience on the volunteer tourist should be examined through the research area of 'sojourner adjustment' rather than the 'psychology of tourism'. This is due to the objective and subjective length of stay on average being greater than for tourists. Therefore we turn to studies exploring the learning and cultural experiences of international students, migrants and workers in foreign cultures.

There have been many studies on the long-term cultural experiences of international student sojourners (cf. Furnham and Bochner, 1982; Kashima and Loh, 2006; Lewthwaite, 1996; Rohrlich and Martin, 1991; Ward and Searle, 1991). In examining the adaptation process of 12 postgraduate students in New Zealand, Lewthwaite (1996) found that the students experienced stress as a result of lack of preparation and language difficulties. There were serious gaps in the communication process. Similarly, Rogers and Ward (1993) note the importance of preparation and therefore examined expectations in the cross-cultural adjustment process. They found that students with realistic expectations experienced a greater ease of adaptation. Additionally, research on pre-service teachers teaching low income minority children has been undertaken to understand initial perceptions and their link with experience. Haberman and Post (1992: 30) found that the teachers used 'these direct experiences to selectively perceive and reinforce their initial preconceptions'. These may be useful studies to understand how miscommunication in cross-cultural experiences can be avoided. In fact, 'this dynamic notion of culture can be a useful framework to guide the interpersonal negotiations needed to transform relations' between host community and volunteer (DePalma, Santos Rego and del Mar Lorenzo Moledo, 2006: 328).

Intercultural exchange through volunteer tourism

Volunteer tourism offers us the opportunity to examine new kinds of cross-cultural relationships in many cases through collaborative goal-orientated projects that take place inside a community's own space. Generally volunteer tourists enter the host space with a valuing of the place and its cultures so in order to understand this we need to have mechanisms that reprioritise the way the tourism experience is constructed. This can take many forms but here we have chosen to focus on how this can occur if the values of those traditionally 'othered' are incorporated into the conceptualisation and research. Providing a volunteer tourism experience where the tourist focus is on the interaction with community members from different cultures has been praised as an effective means to improve the tourism experience and expected to improve cross-cultural sensitivity and reduce the 'othering' of developing countries' cultures.

Recent studies (Griffin, 2004; Raymond and Hall, 2008; Simpson, 2004) have identified the potential of volunteer tourism to foster cross-cultural misunderstanding. They build on Bochner's (1982) 'us' versus 'them' idea of cultural contact. The main argument presented is that these young volunteers are untrained with primarily egoistic motives. Hustinx (2001) argues that this is a new model of volunteering. It is a process of individualisation and self-exploration but in the process, is one of cultural exchange. Volunteer tourism is now a discourse of mutual benefit which involves a two-way process of knowledge sharing to 'reinstate a sense of equality between self and other' (Matthews, 2008: 108).

In volunteer tourism the cross-cultural exchange is a crucial part of the experience. Although the latest topic of interest amongst volunteer tourism academics is the intercultural experience gained by the volunteer (Matthews, 2008; McIntosh and Zahra, 2007; Raymond and Hall, 2008; Spencer, 2008), very few studies site this as an initial motivation for travel. There are two noted exceptions. First, Brown and Lehto (2005) found that cultural immersion was the most important motivation for volunteering amongst much older volunteers. Second, Rehberg's (2005) study of young Swiss adults (average age of 24) resulted in a division of three distinct groups: achieving something positive for others; quest for the new and quest for oneself. The 'quest for the new' involved motives of intercultural exchange including becoming acquainted with the new culture, learning the language and making new friends. Rehberg (2005) recommended that due to this need for cultural immersion, programs should be planned which involved contact and cooperation with the host community.

Intercultural exchange through volunteer tourism can be achieved if it aligns itself with the principles of alternative community-based tourism, sustainability and ecotourism. That is, it should be small scale, not damage the environments on which it depends, aim to empower communities by allowing them to plan and manage the programs, and should foster an understanding between the industry (private and NGOs) and community and between the host and guest (Scheyvens, 2002; Sofield, 2003; Wearing, 2004). The following case study contextualises these ideas and presents the experiences of volunteer tourists working on projects with Youth Challenge Australia.

Case study: Youth Challenge Australia, leading by example

Youth Challenge Australia (YCA) is an NGO operating out of Sydney, Australia, which sends youth between the ages of 18–30 to volunteer in five developing countries as well as having a program in Central Australia. Its parent organisation, Youth Challenge International (YCI), has been operating for 20 years and now has ten partner organisations (including YCA) throughout the South Pacific, Africa and the Americas. The difference between the Youth Challenge model and other volunteer tourism organisations is that the approach to development is through engaging and empowering youth through collaborative development initiatives. This extends beyond youth volunteering in an international context to the involvement of local youth ensuring sustainability in the regions they work. YCA's development focus is driven by projects identified by the local partners and communities and it ensures that:

> As YCA volunteers are hosted by the communities they work with, this deepens links between both the community, and the volunteers themselves. YCA volunteers become part of a unique cross cultural experience that remains in both community members' and volunteers' memories for years to come.
>
> (YCA, 2007)

The experience a YCA volunteer has in another country is only the beginning of their service. The volunteers are actively encouraged to continue their community engagement in their own communities or consider volunteering internationally again.

From 2007–2008 35 YCA volunteers and one YCI volunteer (going to four different countries for six to ten weeks) participated in a quantitative longitudinal study looking at the effects of the volunteer experience on their return home. Of the 36 volunteers who were asked to participate in the study, 33 completed the pre-departure survey and 21 were successfully contacted and therefore completed the return survey. Analysis of the pre-survey data found that the volunteers had a range of initial motivations for undertaking the projects. These results reflect those conducted with other volunteer tourism organisations. That is, the top three motivations were personal development (88 per cent), to discover a different culture/environment (64 per cent) and to help a community (48 per cent).

Further analysis of the data has shown that the volunteers expected to have high to very high exposure to cultural interaction. This expectation is provided by YCA who prepare the volunteers through a weekend workshop as well as a pre-departure information day prior to leaving for their project country. They are prepared for an experience with vastly different outcomes to a traditional tourism experience. Regardless of this, volunteers were concerned with a variety of factors prior to departure. After health and relationships within the group, the next three most concerning factors were language, offending the host community and the community response.[1] Interestingly, the volunteers had little concern for adjusting to new customs (although 34.4 per cent indicated a high to extreme response).

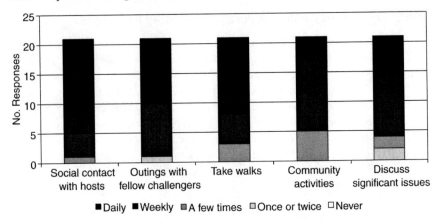

Figure 13.2 How frequently did you undertake the following activities with the host community during your time away?

One aim of the post survey was to understand the cultural experiences that the volunteers had in their project countries. Upon return all the volunteers indicated that they experienced high to very high cultural interaction. Figure 13.2 shows that during the length of the project, the majority of the volunteers had more than weekly contact with the host community where social contact with hosts occurred daily for 76 per cent of the volunteers.

A correlation of the initial level of concern against the experienced level of difficulty of 17 factors found that respondents experienced much lower levels of difficulty than expected with all factors apart from 'language'. This factor had a high expected and even higher experienced level of difficulty. This finding was repeated in another section of the survey looking at six items. Apart from the experienced language proficiency being lower than expected, so too was the knowledge gained from the experience. This suggests that even though the volunteers' experiences were very positive, their adaptation and, consequently, their intercultural exchange with the communities could have been easier and better managed had they been more familiar with the language prior to departure. Although this was found to be slightly lower than expected in the six-item questions, the open-ended comments display a different story. When asked to indicate the knowledge (information, theories) and skills (practical applications) that they gained from the program, 15 of the volunteers gave responses ranging from basic skills to new values. Most of the responses indicated an increased knowledge of the culture and their own adaptation. For example, one respondent noted:

> I learnt a lot about my own capabilities and values. I learnt so much about different cultural interaction, teaching and working with children. I also learnt basic things like cooking in different ways, learning to adapt to certain surroundings and situations, making group decisions and working in a large group.
> (Emma)

As learning is a large part of the volunteer experience noted by the primary motivations found in this and prior studies, these results suggest that it is essential for knowledge sharing to be continually incorporated in the projects. The degree of contact with the host community is very important to the volunteer and is supported by the fact that 38 per cent of YCA volunteers remain in contact with the host community upon return. As a result this relationship between volunteer and host appears to be more equal than that formed in a tourism experience.

Conclusion

It has been acknowledged that presently there is no way of measuring tourist–host contact (Reisinger and Turner, 2003: 68). We suggest that the contact between volunteer and host is very different to tourist and host and therefore requires different attention. In most cases this is due to the nature of the experience, that is, that NGOs are involved with an aim to create empowered communities and issues of equity (Lyons and Wearing, 2008). YCA provides an example of an NGO which has created expectation and learning experiences which have given the volunteers positive experiences. These positive experiences have helped the volunteer to adapt bringing about an intercultural exchange between the volunteer and host community.

Raymond and Hall (2008) note that three components are required for volunteer tourism sending organisations to achieve cross-cultural understanding. The first is that they develop programs that benefit the community, the second is to ensure the programs are recognised as a learning process and not an 'experience', and the third is to facilitate cultural interaction between the volunteer and host community. Although these guidelines are a start we contend that in order to move beyond the traditional demarcations that do not acknowledge how these communities are 'othered' under traditional tourism experiences, a decommodified approach must be taken to achieve the goals of intercultural exchange through volunteer tourism. This will involve empowering the community to take responsibility for the programs and engaging youth in a participatory development role giving them the necessary knowledge to communicate with the host community.

It has been reported in a number of studies that the sheer immaturity of volunteers has in fact created more problems than helped. More often than not the power inequalities still exist and the programs do little to benefit the host community (Simpson, 2004; Sin, 2009). This chapter has explored the programs of one NGO, Youth Challenge Australia, to demonstrate first that volunteer tourism can foster intercultural exchange without introducing the notion of the 'other' simply by making the programs just one step in the volunteer's journey in community service and development. Secondly, when the communities involved in these volunteer tourism programs are given ownership and power over the programs, they create an equitable relationship with the volunteers, one that sees both parties strive to achieve the same goals. YCA provides a model where host communities are not 'othered'. This is because they have empowered the communities who make the decisions as well as provided the volunteer with the necessary knowledge to go through a process of adaptation. According to Intercultural Adaptation Theory, this matching of goals and power equity minimises the cross-cultural misunderstanding and fosters intercultural exchange.

Note

1 These five factors had means of 2.47, 2.63, 2.69, 2.69 and 2.75 on a range of 1 to 5 with 1 being extreme and 5 being none. All other factors (17 in total) were rated neutral or below.

References

Bakhtin, M. M., Emerson, C. and Holquist, M. (1986) *Speech Genres and Other Late Essays* (1st edn), Austin: University of Texas Press.
Bochner, S. (1982) The social psychology of cross-cultural relations. In S. Bochner (ed.) *Cultures in Contact: Studies in Cross-Cultural Interaction*. Oxford: Pergamon Press, pp. 5–44.
Brammer, N., Beech, J. and Burns, P. (2004) Use and abuse of tourism: the Goan experience, *Tourism Culture and Communication*, 5(1): 23–35.
Broad, S. and Jenkins, J. (2008) Gibbons in their midst? Conservation volunteers' motivations at the Gibbon Rehabilitation Project, Phuket, Thailand. In K. Lyons and S. Wearing (eds) *Journeys of Discovery in Volunteer Tourism: International Case Study Perspectives*. Wallingford, UK: CABI, pp. 72–85.
Brohman, J. (1996) New directions in tourism for Third World development. *Annals of Tourism Research*, 23(1): 48–70.
Brown, S. and Lehto, X. (2005) Travelling with a purpose: understanding the motives and benefits of volunteer vacationers. *Current Issues in Tourism*, 8(6): 479–496
Butcher, J. (2005) The moral authority of ecotourism: a critique. *Current Issues in Tourism*, 8(2 and 3): 114–124.
Butcher, J. (2006) Natural capital and the advocacy of ecotourism as sustainable development. *Journal of Sustainable Tourism*, 14(6): 529–544.
Cai, D. A. and Rodriguez, J. I. (1997) Adjusting to cultural differences: the intercultural adaptation model. *Intercultural Communication Studies*, 1(2): 31–42.
Cater, E. (1993) Ecotourism in the Third World: problems for sustainable tourism development. *Tourism Management*, 14(2): 85–90.
Cheong, S.-M. and Miller, M. L. (2000) Power and tourism: a Foucauldian observation. *Annals of Tourism Research*, 27(2): 371–390.
Cunningham, P. (2006) Social valuing for Ogasawara as a place and space among ethnic host. *Tourism Management*, 27(3): 505–516.
DePalma, R., Santos Rego, M. A. and del Mar Lorenzo Moledo, M. (2006) Not just any direct experience will do: recasting the multicultural teaching practicum as active, collaborative and transformative. *Intercultural Education*, 17(4): 327–339.
Dernoi, L. A. (1988) Alternative or community-based tourism. In L. D'Amore and J. Jafari (eds) *Tourism – A Vital Force for Peace*, Vancouver, Canada: D'Amore and Associates, pp. 89–94.
Doron, A. (2005) Encountering the other': pilgrims, tourists and boatmen in the city of Varanasi. *Australian Journal of Anthropology*, 16(2): 157–178.

Ehrichs, L. (2000) *Volunteering in Development : A Post-Modern View.* Available from http://www.worldvolunteerweb.org/browse/countries/lao-pdr/doc/volunteering-in-development.html. Accessed 21 January 2009.

Ellingsworth, H. W. (1983) Adaptive intercultural communication. In W.B. Gudykunst (ed.) *Intercultural Communication Theory: Current Perspectives*, vol. VII, Beverly Hills: Sage Publications.

Foucault, M. (1980) *Power/Knowledge: Selected Interviews and Other Writings 1972–77.* Brighton: UK Harvester.

Foucault, M. (1986) Of other spaces. *Diacritics*, 16(1): 22–27.

Furnham, A. (1984) Tourism and culture shock. *Annals of Tourism Research*, 11(1): 41–57.

Furnham, A. and Bochner, S. (1982) Social difficulty in a foreign culture: an empirical analysis of culture shock. In S. Bochner (ed.) *Cultures in Contact: Studies in Cross-Cultural Interaction*, Oxford: Pergamon Press, pp. 161–198.

Furnham, A. and Bochner, S. (1986) *Culture Shock: Psychological Reactions to Unfamiliar Environments*, London: Methuen.

Galli, R. (1992) Winners and losers in development and antidevelopment theory. In R. Galli (ed.) *Rethinking the Third World: Contributions Towards a New Conceptualization*, New York: Crane Russack, pp. 1–27.

Griffin, T. (2004) *A Discourse Analysis of UK sourced Gap Year Overseas Projects.* Unpublished Masters Thesis, University of the West of England.

Haberman, M. and Post, L. (1992) Does direct experience change education students' perceptions of low-income minority children? *Mid-Western Educational Researcher*, 5(2): 29–31.

Hall, S. (1997) Introduction. In S. Hall (ed.) *Representation: Cultural Representations and Signifying Practices*, London: Sage Publications, pp. 1–11.

Harrison, D. (1992a) International tourism and the less developed countries: the background. In D. Harrison (ed.) *Tourism and the Less Developed Countries*, London: Belhaven Press, pp. 1–18.

Harrison, D. (1992b) *Tourism and the Less Developed Countries*, London: Belhaven Press.

Haywood, K. M. (1988) Responsible and responsive tourism planning in the community. *Tourism Management*, 9(2): 105–118.

Holden, A. (2000) *Environment and Tourism*, London: Routledge.

Hunter, C. and Green, H. (1995) *Tourism and the Environment: A Sustainable Relationship?* London: Routledge.

Hustinx, L. (2001) Individualisation and new styles of youth volunteering: an empirical investigation. *Voluntary Action*, 3(2): 57–76.

Ingram, J. (2009) Volunteer tourism: How do we know it is 'making a difference'? In A. Hergesell and J. J. Liburd (eds) *Proceedings of BEST EN Think Tank IX: The Importance of Values in Sustainable Tourism and First International Symposium on Volunteering and Tourism*, 14–18 June, Singapore, CD Rom, Sydney: University of Technology Sydney.

Jack, G. and Phipps, A. (2005) *Tourism and Intercultural Exchange: Why Tourism Matters*, Clevedon, UK: Channel View Publications.

Jandt, F. E. (2007) *An Introduction to Intercultural Communication: Identities in a Global Community* (5th edn), Thousand Oaks: Sage Publications.

Kashima, E. S. and Loh, E. (2006) International students' acculturation: effects of international, conational, and local ties and need for closure. *International Journal of Intercultural Relations*, 30(4): 471–485.

208 *Stephen Wearing and Simone Grabowski*

Lave, J. and Wenger, E. (1991) *Situated Learning: Legitimate Peripheral Participation*, Cambridge, UK: Cambridge University Press.

Laythorpe, K. (2009) Sustainable living in a Third World country: experiences of long term volunteers in the Kilimanjaro region of Tanzania. In A. Hergesell and J.J. Liburd (eds) *Proceedings of BEST EN Think Tank IX: The Importance of Values in Sustainable Tourism and First International Symposium on Volunteering and Tourism*, 14–18 June, Singapore, CD Rom, Sydney: University of Technology Sydney.

Lewthwaite, M. (1996) A study of international students' perspectives on cross-cultural adaptation. *International Journal for the Advancement of Counselling*, 19(2): 167–185.

Lyons, K. D. and Wearing, S. (2008) Volunteer tourism as alternative tourism: Journeys beyond otherness. In K. Lyons and S. Wearing (eds) *Journeys of Discovery in Volunteer Tourism: International Case Study Perspectives*, Wallingford, UK: CABI, pp. 3–11.

MacCannell, D. (1973) Staged authenticity: arrangements of social space in tourist settings. *American Journal of Sociology*, 79(3): 589–603.

MacCannell, D. (1976) *The Tourist: A New Theory of the Leisure Class*, New York: Schocken Books.

MacCannell, D. (1999) *The Tourist: A New Theory of the Leisure Class*, Berkeley: University of California Press.

MacCannell, D. (2001) Remarks on the commodification of cultures. In V. L. Smith and M. Brent (eds), *Hosts and Guests Revisited: Tourism Issues of the 21st Century*. Elmsford: Cognizant Communication Corporation, pp. 380–390.

Macleod, D. V. L. (2004) *Tourism, Globalisation and Cultural Change: An Island Community Perspective*, Clevedon, UK: Channel View Publications.

Macleod, D. (2006) Cultural commodification and tourism: a very special relationship. *Tourism Culture and Communication*, 6(2): 71–84.

Martinez, T. A. and McMullin, S. L. (2004) Factors affecting decisions to volunteer in nongovernmental organizations. *Environment and Behavior*, 36(1): 112–126.

Mathieson, A. and Wall, G. (1982) *Tourism: Economic, Physical, and Social Impacts*, London: Longman, Wiley.

Matthews, A. (2008) Negotiated selves: exploring the impact of local-global interactions on young volunteer travellers. In K. Lyons and S. Wearing (eds) *Journeys of Discovery in Volunteer Tourism: International Case Study Perspectives*, Wallingford, UK: CABI, pp. 101–117.

McIntosh, A. J. and Zahra, A. (2007) A cultural encounter through volunteer tourism: towards the ideals of sustainable tourism? *Journal of Sustainable Tourism*, 15(5): 541–556.

Mehmet, O. (1995) *Westernising the Third World: The Eurocentricity of Economic Development Theories*, London: Routledge.

Mieczkowski, Z. (1995) *Environmental Issues of Tourism and Recreation*, Lanham: University Press of America.

Mowforth, M. and Munt, I. (2003) *Tourism and Sustainability: Development and New Tourism in the Third World*, (2nd ed. edn), New York: Routledge.

Pearce, P. L. (2005) *Tourist Behaviour: Themes and Conceptual Schemes*, Clevedon, UK: Channel View Publications.

Ponting, J., McDonald, M. and Wearing, S. (2005) De-constructing wonderland: surfing tourism in the Mentawai Islands, Indonesia. *Loisir et Société/Society and Leisure*, 28(1): 141–162.

Prakash, G. (1994) Postcolonial criticism and Indian historiography. In L. Nicholson and S. Seidman (eds), *Social Postmodernism: Beyond Identity Politics*, Cambridge: Cambridge University Press, pp. 87–100.

Raymond, E. M. and Hall, C. M. (2008) The development of cross-cultural (mis) understanding through volunteer tourism. *Journal of Sustainable Tourism*, 16(5): 530–543.

Rehberg, W. (2005) Altruistic individualists: motivations for international volunteering among young adults in Switzerland, *Voluntas*, 16(2): 109–122.

Reisinger, Y. (2009) *International Tourism: Cultures and Behaviour*, Oxford: Butterworth-Heinemann.

Reisinger, Y. and Turner, L. (2003) *Cross-Cultural Behaviour in Tourism: Concepts and Analysis*, Oxford: Butterworth-Heinemann.

Ritzer, G. (2006) Islands of the living dead: the social geography of Mcdonaldization. In G. Ritzer (ed.) *Mcdonaldization: The Reader* (2nd edn), Thousand Oaks CA: Pine Forge Press, pp. 32–40.

Robinson, M. and Boniface, P. (1999) *Tourism and Cultural Conflicts*. Wallingford: CABI Publishing.

Rogers, J. and Ward, C. (1993) Expectation-experience discrepancies and psychological adjustment during cross-cultural re-entry. *International Journal of Intercultural Relations*, 17(2): 185–196.

Rohrlich, B. F. and Martin, J. N. (1991) Host country and re-entry adjustment of student sojourners. *International Journal of Intercultural Relations*, 15(2): 163–182.

Rojek, C. (1997) Indexing, dragging and the social construction of tourist signs. In C. Rojek and J. Urry (eds) *Touring Cultures: Transformations of Travel and Theory*, London: Routledge, pp. 52–74.

Said, E. (1978) *Orientalism*. London: Routledge.

Said, E. (1993) *Culture and Imperialism*. Toronto: Random House.

Scheyvens, R. (2002) *Tourism for Development: Empowering Communities*. Harlow, England: Pearson Education.

Schmidt, H. (1989) What makes development? *Development and Cooperation*, 6: 19–26.

Simpson, K. (2004) Doing development: the gap year, volunteer-tourists and a popular practice of development. *Journal of International Development*, 16(5): 681–692.

Sin, H. L. (2009) Volunteer tourism: 'Involve me and I will learn'? *Annals of Tourism Research*, 36(3): 480–501.

Smith, V. (1989) *Hosts and Guests: The Anthropology of Tourism* (2nd edn), Philadelphia: University of Pennsylvania Press.

Sofield, T. H. B. (1991) Sustainable ethnic tourism in the South Pacific: some principles. *Journal of Tourism Studies*, 2(1): 56–72.

Sofield, T. H. B. (2003) *Empowerment for Sustainable Tourism Development* (1st ed. edn.) Amsterdam: Pergamon.

Spencer, R. (2008) Lessons from Cuba: a volunteer army of ambassadors. In K. Lyons and S. Wearing (eds) *Journeys of Discovery in Volunteer Tourism: International Case Study Perspectives*, Wallingford, UK: CABI, pp. 36–47.

Telfer, D. J. (2002) The evolution of tourism and development theory. In R. Sharpley and D. J. Telfer (eds) *Tourism and Development: Concepts and Issues,* Clevedon, UK: Channel View Publications, pp. 35–78.

Telfer, D. J. (2003) Development issues in destination communities. In S. Singh, D. J. Timothy and R. K. Dowling (eds) *Tourism in Destination Communities*, Wallingford: CAB International, pp. 155–180.

Teo, P. (1994) Assessing socio-cultural impacts: the case of Singapore. *Tourism Management*, 15(2): 126–136.

Tourism Research and Marketing (TRAM) (2008) *Volunteer Tourism: A Global Analysis*, The Netherlands: ATLAS Publications.

Turner, L. and Ash, J. (1976) *The Golden Hordes: International Tourism and the Pleasure Periphery*, New York: St Martin's Press.

Urry, J. (1990) *The Tourist Gaze: Leisure and Travel in Contemporary Societies*, London: Sage.

Valentine, P. S. (1992) Review: nature-based tourism. In B. Weiler and C. M. Hall (eds) *Special Interest Tourism*, London: Belhaven Press, pp. 105–128.

Ward, C., Bochner, S. and Furnham, A. (2001) *The Psychology of Culture Shock*, (2nd edn) East Sussex: Routledge.

Ward, C. and Searle, W. (1991) The impact of value discrepancies and cultural identity on psychological and sociocultural adjustment of sojourners. *International Journal of Intercultural Relations*, 15(2): 209–224.

Wearing, S. (2001) *Volunteer Tourism: Experiences that Make a Difference*, New York: CABI Publishing.

Wearing, S. (2004) Examining best practice in volunteer tourism. In R. A. Stebbins and M. Graham (eds) *Volunteering as Leisure/Leisure as Volunteering: An International Assessment*, Wallingford: CABI Publishing, pp. 209–224.

Wearing, S. and McDonald, M. (2002) The development of community based tourism: re-thinking the relationship between intermediaries and the rural and isolated area communities. *Journal of Sustainable Tourism*, 10(2): 31–45.

Wearing, S., McDonald, M. and Ponting, J. (2005) Building a decommodified research paradigm in tourism: the contribution of NGOs. *Journal of Sustainable Tourism*, 13(5): 424–439.

Wearing, S. and Neil, J. (1999) *Ecotourism: Impacts, Potentials, and Possibilities*, Oxford: Butterworth-Heineman.

Weaver, G. (1987) The process of re-entry. *The Advising Quarterly*, 2 (Fall): 2–7.

Wheeller, B. (1991) Tourism's troubled times: responsible tourism is not the answer. *Tourism Management*, 12(2): 91–96.

Wiarda, H. J. (1988) Toward a nonethnocentric theory of development: alternative conception from the Third World. In C. K. Wilber (ed.) *The Political Economy of Development and Underdevelopment* (4th edn), Toronto: McGraw-Hill, pp. 59–82.

Wilson, P. (1997) Building social capital: A learning agenda for the twenty-first century. *Urban Studies*, 34(5/6): 745–760.

Youth Challenge Australia (2007) *Development Focus.* Available from http://www.youthchallenge.org.au/YCA/DevelopmentFocus. Accessed 30 June 2009.

Yudice, G. (1995) Civil society, consumption, and government in an age of global restructuring: an introduction. *Social Text*, 14(4): 209–230.

14 Volunteer tourism: how do we know it is 'making a difference'?

Joanne Ingram

'Next holiday idea – why not volunteer overseas?'[1]

'Development' has become fashionable: sponsored charity walks, pay per click internet fundraising, 'Fairtrade' shopping, benefit concerts and album recordings all abound. By taking part in everyday acts, people feel they are making a difference. Now tourism has become part of the development trend, particularly through the addition of volunteering travel.

Wearing (2004: 217) describes volunteer tourism as:

> a form of tourism that makes use of holiday-makers who volunteer to fund and work on conservation projects around the world and which aims to provide sustainable alternative travel that can assist in community development, scientific research or ecological restoration.

This form of tourism is burgeoning. Place 'volunteer tourism' in the Google search field and over 300,000 hits greet you.[2] Advertisements for the myriad of volunteer tourism opportunities are selling the promised experience as 'travel with a purpose'. By participating, the volunteer tourist can 'come and make a difference' and 'contribute in a meaningful way'. Yet, what does this mean? To 'make a difference' implies an opportunity to improve the lives of the communities in which the volunteer is working: in essence, to engage with development and poverty alleviation. These are huge claims for a tourism product. So how do we know that volunteer tourism is making a difference? Are these claims in fact nothing more than catchy slogans? Currently, due to a lack of research and the availability of empirical evidence into the impacts of volunteer tourism, it is difficult to assess whether any difference is being made. This chapter argues for a need to explore the impact of volunteer tourism and its relationship with development to assess the claim: volunteer tourism – making a difference. This chapter will begin by examining the concept of volunteer tourism and its complexity in today's globalised world. Attention will then be given to appraising current development theory, and considering whether volunteer tourism has a place in development.

To niche or not to niche? That is the question

Tourism is big business. Since the 1950s international tourism has exploded with worldwide international arrivals increasing almost forty-fold from just over 25 million in 1950 to approximately 903 million in 2007, although it declined in 2009 by 4 per cent to 880 million (UNWTO, 2010). Today it is one of the world's largest industries, accounting for over 11 per cent of the world's gross domestic product (GDP) and employing approximately 200 million people worldwide. A combination of increased mobility, leisure time and disposable income in the industrialised world has facilitated this massive growth, moving tourism from an activity of a privileged few to mass consumerism (Kaplan, 1996; UNWTO, 2008; Wearing, 2001). Accompanying this growth is also an increase in the interests and pursuits of tourists, where 'it seems that nearly every dimension of human culture now has the potential to become a form of tourism' (Sutton and House, n.d.). This has led many tourists to search out alternatives, such as backpacking, ecotourism and gap year travel, to the mass produced tourist products[3] (Callanan and Thomas, 2005). The pressure for operators to provide alternatives and a need to mark difference has led to the emergence of niche markets offering new experiences that accommodate the desire of the postmodern tourist (Wearing, 2001). In a highly competitive and crowded market place, tourism organisations now recognise that niche markets are the means to gaining a competitive advantage. Furthermore, given that mass tourism is viewed increasingly as negatively impacting on the environment and socio-cultural interactions, the World Tourism Organisation (WTO) advocates niche tourism as less intrusive and more likely to offer benefits to host communities when compared to traditional forms of tourism (Robinson and Novelli, 2005).

There has been concern, however, that niche tourism may in fact encourage the spread of mass tourism as large numbers follow in the footsteps of those who have found new and more exciting destinations and environments (Griffin and Boek, 1997). Two examples that demonstrate this are backpacker and trekking tourism. Formerly regarded as niche markets, these two areas have developed into what can legitimately be referred to as 'mass' markets with concomitant negative impacts on host communities. In their relentless search for the 'newest, the most remote, the most exotic' (Biles, Lloyd and Logan, 1999: 220), through the experience of life in a 'foreign' community, backpackers have created enclaves that all but erase these communities. Places such as Khao San Road in Bangkok and Thamel in Kathmandu, for example, are no longer recognisable as 'local', as they have been taken over by the Western backpacker. The recent explosion in trekking tourism has caused comparable concern. Nepal, a leading trekking destination, has witnessed a massive increase in trekkers over the years with an estimated increase of 255 per cent since 1980. Unfortunately, although providing economic prosperity to Nepal's mountain regions, there has been little consideration for the carrying capacity of the fragile environment as trekkers flow into these regions. The inundation of trekkers has created several challenges for indigenous communities, challenges that include mountain paths strewn with litter and eroding from the number of people utilising them, along with deforestation and soil erosion due to an increased demand for wood to construct lodges and fuel fires (Rembert, 2002; Stevens, 2003). For example, in Nepal's Annapurna region, it is estimated that an independent trekker consumes 4.5kg of firewood a day, almost double the daily usage of locals (Nyaupane and Thapa 2006).[4] Backpackers and trekkers are, thus, changing the entire social dynamic of some places as they demand an environment that caters to their needs and imaginations.

Tourism today is far reaching, so much so, that tourism 'has become an intrinsic part of both global and local culture' (Wood, 1997: 1). It is, therefore, not only the economic results, but also the cultural impacts of tourism that should be acknowledged. While niche tourism is generally promoted and understood to provide alternatives that have a minimal impact on the culture of host communities, the examples above demonstrate that this is not necessarily the case, and in fact, there is potential to impact negatively. Similar concern can be raised in relation to volunteer tourism, which can also be categorised as a niche market. Although this form of travel is marketed as having the capacity to 'make a [positive] difference', where is the evidence to support this? This chapter suggests that volunteer tourism can on occasion be yet another tourism product converting 'cultural value' into 'commercial value' (Shepherd, 2002).

Altruistic motivated travel?

In 2004, following the lead of his older brother Prince William,[5] Prince Harry joined the legions of young people participating in gap years and international volunteer tourism projects. During a two-month stay in Lesotho, Southern Africa, Prince Harry worked as a volunteer: participating in building projects, playing with school children and working in the fields (Holman, 2004).[6] These activities by Britain's young royals reflect the massive recent rise in 'gap year' travel. In the United Kingdom alone it has been estimated that in 2003 around 250,000 people aged between 18 and 25 set out for a year's break (Jones, 2004 and Holman, 2004). High profile cases such as those of Prince William and Prince Harry have assisted in the promotion and appeal of the 'gap year' phenomenon:

> If you, like Prince Harry, fancy roughing it for a few months while doing something more useful with your time than getting stoned in Thailand (not that we're ever suggesting that Prince Harry would do that!) then SPW [Student Partnership Worldwide] might be the answer. SPW's gap year schemes offer young people the opportunity to spend up to a year in countries across the Third World.
>
> (Holman, 2004)

Promotions, like this one, are capturing the imagination of today's youth, and it is this market in particular that is the target of organisations offering volunteer tourism projects. Having said this, adults of all ages are enthusiastic participants in this niche market.

Volunteering can be defined as 'un-coerced help offered ... with no or, at most, token pay done for the benefit of both the people and the volunteer' (Brown, 2005: 483). It can also be viewed as 'a form of civic engagement through which individuals can make meaningful contributions to *their own visions* of societal well-being' (Brown, 1999: 3). In recent years we have seen the growth of an alternative tourism product, volunteer tourism, which combines this 'un-coerced help' with some form of leisure travel. The combination of work and leisure in volunteer tourism suggests a contradiction and the likelihood of tensions, particularly as the two appear to be complete opposites. Yet people have warmed to the product, even when it relies on the undertaking of unpaid work whilst on holiday. Why is this? The answer may lie within the complexities of the postmodern world.

The blurring of boundaries: when is it work? When is it leisure?

A characteristic of postmodernism is that it blurs and dissolves boundaries. Such blurring has created a world of fragmented identities and anxiety, as people lack a clear sense of whom they are, and what their place is in the world. This anxiety is increasing ego-centricity and individualism in the West, and is fuelling people's desires to search out an authenticity they feel is missing in their lives (MacCannell, 1992). The demands and constraints on modern societies are immense and, in general, create an environment that works for individuals and against family cohesion. The opportunities or risks that had previously been predetermined by family, class or community now require interpretation and processing by individuals: 'the ethic of individual self fulfilment and achievement is [thus] the most powerful current in modern society' (Beck and Beck-Gernsheim, 2002: 22). This need to search out 'truths' to override individual insecurities has lead to a growth in tourism linked to pre-modern cultures and the 'discovery' of places imagined to be 'untouched' by modern society. Seeking out primitive places in the world and experiencing the 'whole', that is, people and geographical place, is a way people are soothing their feelings of fragmentation (Blom, 2000; MacCannell, 1992). Such feelings can, in part, be attributed to the stresses and demands of the workplace.

Today the boundary between professional and personal spaces has become vague as globalisation and deregulation have transformed the way people work. Flexible working arrangements and an 'individualisation of labour', organised around integrated communication systems, have become workplace standards in the West (Castells, 2000). Coinciding with these standards is the accepted culture of long working hours, a culture that is being fuelled by a combination of: the changed nature of people's careers, downsizing and the subsequent increase in people's workloads, and advances in technology that have enabled people to work outside the traditional office space and the regular pattern of nine to five, Monday to Friday. The notion of work/life balance has become a buzzword yet it is increasingly difficult to achieve as people are spending long hours at work, as well as having contact with the workplace on their holidays. Advances in technology have moved the office to the beach, ski resort and cruise liner, creating an anticipation that, because it is now easy to keep in touch, workers should do so (Green, 2007; Paton, 2006). These changes have culminated in an expectation that, today, all time should be productive (Ng and Feldman, 2008). It can be argued, therefore, that the boundary between work and leisure is becoming indistinguishable, and in the form of volunteering and tourism, the combination of work and leisure is not a contradiction, but a rational match.

Altruism or ego-centric?

Although travelling overseas as a part of modern volunteerism has origins dating back to the early twentieth century, the niche market of volunteer tourism is considered a recent trend, whereby the focus has shifted away from volunteerism in the direction of tourism (Wearing, 2004). Research relating to the volunteer tourism field is currently limited. Much of the literature and research to date focuses on the volunteer, their motives and outcomes, and highlights the prospect of individual growth through participation in volunteer projects. In particular, findings of the research suggest that participation provides individuals with an opportunity to re-evaluate their core beliefs and values, as well as the intrinsic rewards gained from the volunteering experience (Coghlan, 2006; Wearing, 2001).

Sally Brown's (2005) study into why people volunteer on their holidays reveals that a strong motivational factor is the opportunity the experience provides in being able to immerse oneself in the 'authenticity' of a place, experiencing the local culture and interacting with people from elsewhere. Brown's findings support the assertion that 'volunteer tourism [has been] recast within the context of postmodern tourism and the ... attractiveness of the "other"' (Coghlan, 2008: 184). Such motives demonstrate that although volunteer tourism is considered a form of altruistically motivated tourism, there is an element of ego-centricity, as volunteers seek cultural engagements and a variety of experiences from their participation (McIntosh and Zahra, 2007). 'The volunteer vacation phenomenon, [therefore] appear[s] to bridge the altruistic motives of volunteering with the general commodified tourism experience' (Brown, 2005: 494).

It is clear, therefore, that considerable research time has been devoted to those tourists who feel motivated, for whatever reason, to volunteer. Volunteer tourism, by definition, however, requires a community to which the volunteer 'contributes'. Yet, what is known of the response of these communities to the input from volunteer tourism? Extensive searches of the literature reveal, in fact, that almost no evidence exists of studies examining, from any perspective, the effect volunteer tourism has on host communities. Part of the broader project in which this chapter resides is consideration of the suggestion that the myriad of advertisements laying claim to the volunteer tourism experience as 'contributing in a meaningful way' may be nothing more than catchy slogans. A lack of focus on the hosts, their needs, and the impacts and outcomes for those both directly involved in volunteer tourism projects, as well as of the wider community, suggest this is a possibility. Participants and scholars alike have overlooked a vital component of the tourism/volunteering interaction. Such an omission could be viewed as a demonstration of post-colonial 'othering', whereby the 'other' is of secondary importance to the volunteer's reading of the experience. Is volunteer tourism, in fact, objectifying the members of host communities by failing to acknowledge their agency? Do volunteers and the organisations offering volunteer tourism experiences truly seek to 'make a difference'? In seeking to 'make a difference' do the volunteers and the organisations offering volunteer tourism do so without really understanding the needs of the communities they seek to assist?

It is likely that volunteer tourism attracts a wide spectrum of volunteers, ranging from those where altruistic motives are paramount, to those whose own interests dominate. As a niche tourism product, volunteer tourism combines consumption and participation. The volunteer is seeking an experience that blends a leisure component with, on some level, an opportunity to assist others. Even with the best of intentions, however, the volunteer's contributions are prone to be influenced by 'their own visions' of how society should be. As an example of the postmodern pastiche, where like work and leisure, egoism and altruism fuse together, can volunteer tourism reasonably expect to make a significant difference to the lives of the communities it claims to assist (Mustonen, 2007)?

Helping is not enough

The ethical concept that all humans are equal does not equate in today's world, in the sense that inequality and poverty are prevalent throughout (Singer, 2002). Globalisation policies have driven many people into abject poverty, which in turn has lead to those from wealthier societies putting their hands up to help. However, as the film 'The Good Woman of Bangkok', by Australian documentary-maker Dennis O'Rourke, demonstrates, wanting to assist is not necessarily enough. This film follows Aoi, originally from a poor peasant village in Thailand, now working as a prostitute in Bangkok. O'Rouke meets Aoi and over a period of nine months engages in a complex relationship as her client, lover and director. Wanting to 'rescue' Aoi from her seedy life, the director offers to buy Aoi and her family a rice farm. Although Aoi initially moves to the farm, the epilogue of the film shows Aoi back working the streets of Bangkok (Hinrichsen, n.d.; O'Rouke and Rowe, 1991):

> I bought a rice farm for Aoi and I left Thailand. One year later I came back but she was not there. I found her working in Bangkok in a sleazy massage parlour called 'The Happy House'. I asked her why and she said, 'it is my fate'.
> (Williams, 1997: 84)

The release of this film in the early 1990s provoked controversy over O'Rouke's use of tensions between a heterosexual white Western male and an 'Asian, third world, female "other"', and was condemned as 'an act of exploitative hypocrisy, of selfish posturing wrapped in bleeding heart white liberal guilt' (Ang, 1997: 2). It does, however, highlight the complexities of attempting to draw people out of poverty and harm, and how a lack of understanding can, even with good intentions, ultimately lead to negative outcomes. O'Rouke failed to understand Aoi's situation and the pressures compelling her to live the way she did. Removing Aoi from her life in Bangkok was merely addressing a symptom rather than peeling away the layers to locate the root cause of her circumstances. Yet, locating the root cause of a problem is merely the start, resolving the problem is often more difficult. Wanting to help is not always enough. Today, theories of development advocate the importance of working alongside people to isolate the root causes of their problems and to identify possible solutions.

Working in partnership

Development discourse is a product of the West, evolving from modernisation theory that espoused the superiority of the 'developed West' over the 'undeveloped Third World'. This theory placed the 'Third World' in a position of dependency, and considered that progress was only possible via the utilisation of the capacities and knowledge of the West (Scheyvens, 2002). Current development thinking has gradually evolved from this position.

A report by the South Commission defines development as:

> a process which enables human beings to realise their potential, build self-confidence, and lead lives of dignity and fulfilment. It is a process which frees people from the fear of want and exploitation. It is a movement away from political, economic, or social oppression.
>
> (South Commission, 1990)

Non-governmental organisations (NGOs), particularly over the past two decades, have played a vital role in assisting in the development of communities. This in turn has led to a massive growth of international NGOs, recorded as having increased from 832 operating in 1951 to an estimated 40,000 today. Considered the answer to resolving many societal problems, the capacities in which NGOs seek to build are wide ranging: social, political, cultural, technical, financial, to mention a few (Ossewaarde, Nijhof and Heyse, 2008). Building capacities, however, is not easy; yet, NGOs are increasingly working in partnership with local communities to achieve this objective.

A weakness that has been noted in development organisations is what is referred to as the 'salt and pepper syndrome', stressing that assistance achieves little if it is scattered randomly like salt and pepper. To be effective, participation rather needs to be concentrated and carried out through long-term partnership (Eade, 2007; Poulton, 1988). Advocates of the need to work in partnership note that this process requires mutual respect, trust and accountability. In recent years there has been a move in development theory from a top-down, macro approach to one of partnership and participation at a micro level. If capacity building is truly the objective, however, then NGOs need to be willing to alter their practices as required, so as to work alongside their partner. In particular, as Eade (2007: 637) notes, a cooperative approach means NGOs need to 'get ... out of the driving seat and learn … to trust their chosen partner's navigational skills. Just because they paid to fill up the tank does not give NGOs the right to determine the route'. NGOs, therefore, need to analyse and reflect on their role and actions. As the saying goes 'if you give a man a fish, you feed him for a day, if you teach him to fish, you feed him for a lifetime', but as Eade (2007: 634) points out, 'What if that fisher is not a man but a woman? And what if she doesn't own the water in which she is fishing? … [and] what if the NGO does not even know how to fish?'. NGOs have a responsibility to continually review and reflect on their role so that they are vigilant of any changes occurring within the communities and surrounds in which they are operating (Eade, 2007).

The World Bank identifies four levels of participation within the development paradigm: information sharing, consultation, decision making and initiating action (Paul, 1987). Each level is signified by an increase in the intensity of participation and by differences in the relationship between the outside agency and local beneficiaries. The lower level participation is whereby the outside agency 'shares information' with the local beneficiaries in relation to the project being undertaken. This level is very much a top-down approach and allows little input from the community being assisted. The 'consultation' level is still a top-down approach, however, because information flows are more equal, the outside agency factors in local knowledge when planning and implementing project(s). At the 'decision making' level, local communities do have some influence over projects, however it is the final stage, 'initiating action', which provides for the greatest participation from local beneficiaries. Under this level, information and influence over projects are largely a bottom-up approach with outside agencies retaining minimal control (Lane, 1995; Paul, 1987). A review cited by Lane (1995) reveals that of various development projects examined, the most common participation levels are those at the lower levels whereby outside agencies retain control over decision making and the resources involved. The common process at this stage looks to be more of an 'add on' to existing operations, than an organic one (Lane, 1995). It appears, therefore, that participation in development has a long way to go in most cases before it reaches the participative level whereby locals are initiating actions; or whereby '"we" [are] participat[ing] in "their" project, not "they" in "ours"' (Chambers, 1995: 30).

So where does this leave volunteer tourism? If the most effective approach advocated in development theory today is one that encourages community needs to be determined locally and met by working alongside long-term partners, can this tourism initiative truly help to make a difference?

As Kate Simpson (2005) points out, the publicity literature of organisations offering volunteer experiences fails to identify the specific needs of the host community that the volunteers will assist in meeting, rather, preferring to create a space where the volunteer is vital to the host community (Simpson, 2005). Furthermore, it is difficult to locate long-term strategic planning of projects within volunteer tourism organisations. As such, volunteer tourism appears to ignore the root causes of poverty and inequality. Instead, volunteer tourism propagates a public myth of development, one of simplicity, where participation and good intentions are considered enough and the use of unskilled labour is validated as the 'solution' (Simpson, 2004). The increasing popularity of volunteer tourism, thus, makes the '"face" of development ... both highly visible and consumable' (Simpson, 2004: 682).

Volunteer tourism reduces development to an act of 'doing'. The current volunteer tourism model externalises development, whereby 'development is based *outside* of stakeholder communities' (Simpson, 2004: 685). This model is in opposition to the participatory approaches advocated by development practitioners currently working in the field (Simpson, 2004). In volunteer tourism's concept of development, the volunteers are central. In fact, based on research currently available, we could be led to believe that the volunteer is the only entity of importance. Host communities are passive in the equation, portrayed as the Third World 'other', in desperate need of the volunteers' assistance. Volunteer tourism, thus, objectifies 'the other'. Such portrayals are perpetuating colonial and imperialistic attitudes of 'them' and 'us', which have a danger of creating power imbalances between the volunteers and host communities (Cheong and Miller, 2000). It is difficult at this stage, therefore, to determine any genuine value of volunteer tourism's contribution towards 'making a difference'.

Conclusion

Volunteer tourism is an innovative concept that combines both travel and a social conscience. The concept is a complex one, and because of this, it is difficult to assess the volunteer tourism market as a whole. As with any initiative that aims to assist people in capacity building, poverty alleviation and development, it needs to follow good development principles. The evidence available at present suggests otherwise, and therefore, the verdict as to whether volunteer tourism is making a difference is currently open. The extent to which volunteer tourism follows good development principles is largely dependent on the type of organisation and the projects on offer. At its best, volunteer tourism has the potential to offer benefits to both the host community and the volunteer. Through volunteer tourism projects, volunteers can assist in meeting local needs (like building projects and teaching English), whilst for the volunteer it is an opportunity for self-development and enrichment of the CV. Alternatively, at its worst, volunteer tourism may have negative impacts similar to other niche markets (like the trekking and backpacker markets discussed). Volunteer tourism commodifies poverty and has the potential of creating power imbalances between the host and volunteer. Volunteers can view themselves as superior, there to offer the 'backward' poor their assistance and to pass on their knowledge as 'they know best'. This attitude can be viewed as colonial and imperialistic, hence leading to a 'one sided domination and exploitation of members of visited societies by the privileged classes' (Cheong and Miller, 2000: 371–372). As discussed, wanting to help is not necessarily enough. The complexity of volunteer tourism, therefore, requires that the merits of each project be assessed to determine the impacts of volunteer tourism and whether it is making a difference. It is likely, that for volunteer tourism to have a place in development, the industry needs to proactively review their processes and more closely align them to those advocated by current development theories. Much like NGOs, volunteer tourism organisations have a responsibility towards the host communities in which they are working alongside. Currently there is a failure to acknowledge the value of host community agency. Without extensive research into the impacts of volunteer tourism on host communities, therefore, the niche market of volunteer tourism will find it difficult to substantiate its claims of 'making a [positive] difference', instead continuing to be viewed with scepticism as a variety of neo-colonialism rather than innovative development.

Notes

1   STA Travel Academic Newsletter, September 2008: 'Next holiday idea – why not volunteer overseas?'
2   305,000 hits. Accessed 22 April 2009, from Google Search (http://www.google.com.au).
3   Gap year definition: 'a period of time taken by a student to travel or work, often after high school or before starting graduate school, as a break from formal education' (http://dictionary.reference.com/browse/gap year). Gap years can involve many activities though they generally incorporate an element of travel as well as periods of work, either voluntary or paid.
4   Estimated daily usage per person 2.5kg (Nyaupane and Thapa, 2006).
5   Prince William participated in gap year travel in 2000, including voluntary work in Chile.
6   Seeking to highlight the plight of people in Lesotho, particularly the number of children orphaned because of Aids, Prince Harry also devised a television documentary during his visit, 'The Forgotten Kingdom: Prince Harry in Lesotho', which raised over £500,000 for a fund in Lesotho

References

Ang, I. (1997) Foreword. In C. Berry, A. Hamilton and L. Jayamanne (eds) *The Filmmaker and the Prostitute: Dennis O'Rouke's The Good Woman of Bangkok*, Sydney: Power Publications, pp. 1–3.
Beck, U. and Beck-Gernsheim, E. (2002) *Individualisation*. London and Thousand Oaks: Sage Publications.
Biles, A., Lloyd, K. and Logan, W. S. (1999) Romancing Vietnam: the formation and function of tourist images of Vietnam. In J. Forshee, C. Fink and S. Cate (eds) *Converging Interests: Traders, Travelers, and Tourists in Southeast Asia*. Berkeley: University of California Press, pp. 207–234.
Blom, T. (2000) Morbid Tourism – a postmodern market niche with an example from Althorp. *Norwegian Journal of Geography*, 54: 29–36.
Brown, E. (1999) Assessing the value of volunteer activity. *Nonprofit and Voluntary Sector Quarterly*, 28(3): 3–17.
Brown, S. (2005) Travelling with a purpose: understanding the motives and benefits of volunteer vacationers. *Current Issues in Tourism*, 8 (6): 479–496.
Callanan, M. and Thomas, S. (2005) Volunteer Tourism – deconstructing volunteer activities within a dynamic environment. In M. Novelli (ed.), *Niche Tourism: Contemporary Issues, Trends and Cases*. Oxford and Burlington: Elsevier Butterworth Heinemann, pp. 183–200.

Castells, M. (2000) Materials for an exploratory theory of the network society. *British Journal of Sociology*, 51(1): 5–24.

Chambers, R. (1995) Paradigm shifts and the practice of participatory research and development. In N. Nelson and S. Wright (eds) *Power and Participatory Development: Theory and Practice*. London: ITDG Publishing, pp. 30–42.

Cheong, S. and Miller, M. L. (2000). Power and tourism: a Foucauldian observation. *Annals of Tourism Research*, 27(2): 371–390.

Coghlan, A. (2006) Volunteer tourism as an emerging trend or an expansion of ecotourism? A look at potential clients' perceptions of volunteer tourism organizations. *International Journal of Non-profit and Voluntary Sector Marketing*, 11: 225–237.

Coghlan, A. (2008) Exploring the role of expedition staff in volunteer tourism. *International Journal of Tourism Research*, 10: 183–191.

Eade, D. (2007) Capacity building: who builds whose capacity? *Development in Practice*, 17(4): 630–639.

Green, E. (2007) Many taking work with them on vacation. *Deseret News (Salt Lake City)*, August 26. Accessed 10 March 2009 from http://findarticles.com/p/articles/mi_qn4188/is_20070826/ai_n19490663.

Griffin, T. and Boek, N. (1997) Alternative paths to sustainable tourism: problems, prospects, panaceas and pipe dreams. In F. M. Go and C. L. Jenkins (eds) *Tourism and Economic Development in Asia and Australasia*. London: Pinter, pp. 321–337.

Hinrichsen, L. (n.d.) The Good Woman of Bangkok. *Encyclopaedia of the Documentary Film*. Available from http://cw.routledge.com/ref/documentary/bangkok.html. Accessed 20 March 2009.

Holman, R. (2004) *Mind the Gap*. Available from http://www.positivenation.co.uk/issue101/features/feature3/feature3.htm. Accessed 7 October 2008.

Jones, A. (2004). Department of Education and Skills Research Report No. 555. *Review of Gap Year Provision*. London: University of London.

Kaplan, C. (1996) *Questions of Travel: Postmodern Discourses of Displacement*. Durham and London: Duke University Press.

Lane, J. (1995) Non-governmental organisations and participatory development: the concept in theory versus the concept in practice. In N. Nelson and S. Wright (eds) *Power and Participatory Development: Theory and Practice*. London: ITDG Publishing, pp. 181–191.

MacCannell, D. (1992) *Empty Meeting Grounds: The Tourist Papers*. London and New York: Routledge.

McIntosh, A. and Zahra, A. (2007) A cultural encounter through volunteer tourism: towards the ideals of sustainable tourism? *Journal of Sustainable Tourism*, 15(5): 541–556.

Mustonen, P. (2007) Volunteer tourism – altruism or mere tourism? *Anatolia: An International Journal of Tourism and Hospitality Research*, 18(1): 97–115.

Ng, T. W. H. and Feldman, D. C. (2008) Long work hours: a social identity perspective on meta-analysis data. *Journal of Organizational Behaviour*, 29: 853–880.

Nyaupane, G. P. and Thapa, B. (2006) Perceptions of environment impacts of tourism: a case study at ACAP, Nepal. *International Journal of Sustainable Development and World Ecology*, 13: 51–61.

O'Rourke, D. and Rowe, G. (Producers), and O'Rourke, D. (Director) (1991). *The Good Woman of Bangkok* [Film]. Australia.

Ossewaarde, R., Nijhof, A. and Heyse, L. (2008) Dynamics of NGO legitimacy: how organising betrays core missions of INGOs. *Public Administration and Development*, 28: 42–53.

Paton, N. (2006) Workers struggle to escape work while on holiday. Available from http://www.management-issues.com/2006/8/24/research/workers-struggle-to-escape-work-while-on-holiday.asp. Accessed 10 March 2009.

Paul, S. (1987) World Bank Discussion Papers 6. *Community Participation in Development Projects: The World Bank Experience.* Washington: The World Bank.

Poulton, R. (1988) On theories and strategies. In R. Poulton and M. Harris (eds), *Putting People First: Voluntary Organisations and Third World Organisations.* London and Basingstoke: Macmillan Publishers, pp. 11–32.

Rembert, T. C. (2002) Going green: walkabout; walking tours are a great way to see the world, at your own pace. *The Environment Magazine,* 13(4): 46.

Robinson, M. and Novelli, M. (2005) Niche tourism: an introduction. In M. Novelli (ed.) *Niche Tourism: Contemporary Issues, Trends and Cases,* Oxford and Burlington: Elsevier Butterworth Heinemann, pp. 1–11.

Scheyvens, R. (2002) *Tourism for Development: Empowering Communities.* Harlow (UK): Pearson Education Ltd.

Shepherd, R. (2002) Commodification, culture and tourism. *Tourist Studies,* 2(2): 183–201.

Simpson, K. (2004) 'Doing development': the gap year, volunteer-tourists and a popular practice of development. *Journal of International Development,* 16: 681–692.

Simpson, K. (2005). Dropping out or signing up? The professionalisation of youth travel. *Antipode,* 37(3): 447–169.

Singer, P. (2002). *One World: The Ethics of Globalisation.* Melbourne: The Text Publishing Coy.

South Commission (1990) *The Challenge to the South: The Report of the South Commission.* Oxford: Oxford University Press.

Stevens, S. (2003). Tourism and deforestation in the Mt Everest region of Nepal. *The Geographical Journal,* 169(3): 255–277.

Sutton, P. and House, J. (n.d.) *The New Age of Tourism: Postmodern Tourism for Postmodern People?* Accessed 7 October 2008 from http://www.arasite.org/pspage2.htm.

UNWTO (2008) Quick overview of key trends. *World Tourism Barometer,* 6(2), 1–55.

UNWTO (2010) Quick overview of key trends. *World Tourism Barometer,* 8(1), 1–59.

Wearing, S. (2001) *Volunteer Tourism: Experiences that Make a Difference.* Wallingford and Cambridge (USA): CABI Publishing.

Wearing, S. (2004) Examining best practice in volunteer tourism. In R. A. Stebbins and M. Graham (eds), *Volunteering as Leisure/Leisure as Volunteering: An International Assessment.* Wallingford: CABI Publishing, pp. 209–224.

Williams, L. (1997) The ethics of documentary intervention: Dennis O'Rouke's *The Good Woman of Bangkok.* In C. Berry, A. Hamilton, and L. Jayamanne (eds), *The Filmmaker and the Prostitute: Dennis O'Rourke's The Good Woman of Bangkok.* Sydney: Power Publications, pp. 79–90.

Wood, R. E. (1997) Tourism and the state: ethnic options and constructions of otherness. In M. Picard and R. E. Wood (eds), *Tourism, Ethnicity and the State in Asian and Pacific Societies.* Honolulu: University of Hawaii Press, pp. 1–34.

15 How does it make a difference? Towards 'accreditation' of the development impact of volunteer tourism

Liam Fee and Anna Mdee

Introduction

Whilst some argue that volunteer tourism is nothing more than neo-colonialism, we propose that it can (under certain conditions) make a positive contribution to local communities in developing countries and can also contribute to a 'globalising, humanising civil society'. We also argue that an increase in volunteer tourism is likely to be an unstoppable trend as international travel and easy global communication make 'do-it-yourself' development activities ever more possible. In this chapter, we consider further the conditions required for volunteers to have a positive rather than a negative or neutral impact.

At present the volunteer tourism industry is relatively unregulated and previous work (Mdee and Emmott, 2008) explored the potential for some kind of fair trade labelling in order to curb some very poor practices and profiteering.

This chapter explores this idea in more detail and considers the theoretical and practical implications of three possible formats for regulation:

- fair trade labelling (external audit)
- a 'good for development' label as proposed by DFID and others
- a membership organisation with a self-certification audit.

The discussion draws on research conducted into recent debates in the sector and an attempt by a small UK NGO, which organises volunteer travel to Tanzania, to create a Responsible Volunteering Association (ReVA).

Volunteer tourism: who does it benefit?

There are a great number of organisations who send volunteers to developing countries to work on projects. Many of these target the 'gap year' market; young people wishing to combine travel, adventure and an ethical volunteering experience before or immediately after university. The value of this market was estimated to be £5 billion in 2005 and, somewhat remarkably, it is predicted to rise to £20 billion by 2010 (Mdee and Emmott, 2008). There are 800 or more organisations offering volunteering placements to over 200 countries, with 100,000 young

British people undertaking such placements every year (Jones, 2005). Involved in this marketplace are both profit and not-for-profit organisations. I-to-I is an example of an organisation which shows the commercial profitability of volunteer tourism through their sale to First Choice Holidays for roughly £20 million in early 2007 (Mdee and Emmott, 2008).

In order to develop a mechanism to account for and encourage good practice in relation to volunteer tourism, it is necessary to first try to understand the nature of its impact. The central question for such a mechanism must be: can short-term volunteering have a positive social, economic or cultural benefit in developing countries? This chapter will consider the small but growing critical literature on the subject in which three positions can be identified: first, that short-term volunteering is neo-colonialist; second, that short-term volunteering can contribute towards a 'humanising global civil society'; or third, that it has little or no impact. These views are counter-balanced by the marketing claims of volunteer providers who are necessarily compelled to make rhetorical claims concerning the positive and transformative impact of volunteer tourism on the 'poor'.

Neo-colonialism: stereotypes and myths

The charge of 'neo-colonialism' is an often propagated argument in volunteer tourism, development and globalisation literature. However, understanding what it is and how it relates to volunteer tourism is not a simple matter. If the question 'how can volunteering in developing countries be viewed as neo-colonialist?' is asked, it seems far-fetched to suggest that today's largely young volunteers can be placed in the same category as Victorian era colonisers. However, to adopt this view would be an over-simplification, and the nature and definition of neo-colonialism must be considered further.

Kwame Nkrumah argued that 'The essence of neo-colonialism is that the State which is subject to it is, in theory, independent and has all the outward trappings of international sovereignty. In reality its economic system and thus its political policy is directed from outside' (1965: 1). In today's society, neo-colonialism can be seen as the domination of European discourses (such as history, philosophy and 'development') and of the European and American forces of capitalism (McEwan, 2008: 124–125). On this basis, and before the impact of volunteer tourism can be considered, the post-colonial critique must be reviewed. If we view post-colonialism to be 'ways of criticizing the material and discursive legacies of colonialism' (Radcliffe, 2005: 84) then the language of development and volunteer tourism providers can be seen to be neo-colonial. If we take, for example, the term 'third world', peoples, societies and cultures in the 'third world' are conveniently homogenised into one group – one with a un-developed economy and a simplistic or corrupt political system. Moreover, according to the post-colonial critique, it creates an 'us and them' or 'self and other' mentality (McEwan, 2008: 125).

Simpson suggests that short-term volunteer tourism providers, or 'the gap year industry' 'presents itself as operating, a geography of homogenous peoples, a geography without history or politics' (Simpson, 2004: 683). On an objective level, then, is this correct? Does the gap year industry have a simplistic, neo-colonialist, 'us and them' view of 'the third world'? What follows are some statements found on the websites of some profit-making volunteer tourism providers: 'in East Africa, the people are friendly, travel is relatively inexpensive and everyone speaks English', 'a life-changing opportunity to help develop underprivileged children's skills and equip them for a brighter future in further education' and, perhaps best of all, 'It has a vibrant cultural life and though its simple charms attract more than 1 million visitors every year, it still manages to keep its relaxed way of life'[1] (i2i, 2009, n.p.).

What we see here are three examples, all from i2i, a sub-set of First Choice Holidays, of Simpson's critique. The last quote, for example, defines the geographic space Simpson describes. Furthermore, it homogenises Costa Rica as 'relaxed' and even suggests that, despite the influx of (presumably) Western tourists, Costa Rica is a still a little backward: 'Soak up jaw-dropping ancient culture, vibrant colours and the many flavours of Asia whilst contributing to vital community and conservation projects helping this beautiful country recover from a turbulent past' (the Leap, 2009).

The Leap, another private firm in the volunteer tourism market also appear to fall into the neo-colonial trap, as indeed do Realgap:

> As in many developing countries, Ghana has a large number of abandoned children and orphans. The care and attention needed by these children is often not received due to lack of staff and government funding. Volunteers are able to make a significant difference to the lives of orphans by offering them the love and attention that so many are missing.
>
> (Realgap, 2009: n.p.)

While the quotes above do not, and were never designed to, prove the neo-colonial *intentions* of volunteer tourism providers, they do back up Simpson's (2004) argument in that certain stereotypes and myths of the 'undeveloped' are reproduced as market 'hooks' to draw in customers. Of course the volunteer tourism industry is not a lone offender in this, and perhaps is only reflecting the same myths of the 'undeveloped' held by many individuals and the popular media in 'developed' countries. What they do invoke is the suggestion that a volunteer (irrespective of their skills) can make a difference to poverty and deprivation. It is the reinforcement of this myth that is perhaps one of the most problematic aspects of the volunteer tourist encounter and contributes to an over-expectation on the volunteer's part of their own direct impact, and a seeming disregard for a need to learn about and respect other ways of doing and being. This is supported by the direct experience of the authors as Trustees of a UK charity which organises such placements.

In addition to the neo-colonialist argument is the assertion that volunteer tourism has no or negative impact on the host community. To this end Voluntary Service Overseas (VSO) has argued that many volunteer tourism schemes are 'badly planned and spurious schemes ... which satisfy the demands of students rather than the needs of locals' (VSO, 2007). As Brown and Hall point out 'volunteers ... often have little experience or knowledge of the work they are undertaking' (2008: 845). There are many examples of harm or deception being caused to both volunteer and host community. For example, the 'turtle census' in South America, which repeatedly counts and surveys the same groups of animals (Baldwin, 2007: 15), achieves nothing for either party. In another example, a classroom in Tanzania was repeatedly painted by UK school groups with the local community smearing the freshly painted walls with mud between visits.[2] It is certainly true that existing attempts to improve practice do not sufficiently consider the host community but instead focus on the consumer experience of the volunteer. The activities of Tourism Concern and the Year Out Group on volunteer tourism are examples of this, although Tourism Concern do acknowledge the gap in their work (Power, 2007).

Whilst the neo-colonial and negative impact critiques do have some validity, they do not account for the potentially positive aspects of volunteer tourism. In earlier work, Mdee has argued that under certain conditions volunteer tourism can be part of a 'globalising, humanising civil society' (Mdee and Emmott, 2008). VSO's dismissive view smacks of elitism that seeks to keep volunteering only for those who are considered 'fit' for the field but this is impractical and illogical in a globalised and networked world. Instead, we do not wish to argue that it is the individual volunteer who makes a 'difference' but rather the collective process of cultural exchange that volunteering facilitates. In addition, if volunteer tourism is considered as part of tourism per se, then the benefits are clearer. If we take Tanzania as an example, the Tanzania tourism sector has contributed to strong economic growth over the last decade; however, in order for the growth of the sector to impact on poverty then the leakages from the Tanzania economy (to pay for imports of goods and services for the sector) need to be minimised (Kweka *et al.*, 2003). Volunteer tourism in common with other forms of community-based tourism can, when practised responsibly, have significant economic benefits for host communities (Nelson, 2004; Goodwin, 2009).

The distribution of benefits from volunteer tourism

The purpose of this chapter is not to join the volunteer tourism critic discourse per se, but rather to recognise its problems and imperfections and to understand impact in a broader way. Therefore, in this section we will outline the potential nature of the impact of volunteer tourism on both the volunteer and on the host community. Defining the substance of a potentially positive impact will later allow us to outline a system whereby volunteer tourist agents might account for their own practice.

One possible line of differentiation that may link to impact is the constitution of the volunteer tourism agent. Simpson's (2004) criticisms of volunteer tourism are aimed at private companies whilst she offers praise to Student Partnerships Worldwide (SPW), a charity. At the heart of this distinction between private and charity/social enterprise providers is motivation, the fundamental difference between private profit driven on one side, and 'development' driven on the other. As Power identifies, 'At one end of the spectrum you have not-for-profit international development organisations using volunteers to accomplish the work they undertake, at the other there are commercial bodies offering placements to large numbers of volunteers and in many different countries' (2007: 29–30). By extension it would then follow that the placements, or 'products', offered by these types of organisations are also very different, at least in terms of their impact on the local community. To some extent, this argument has some merit. A small local NGO that hosts paying volunteers who have relevant skills is likely to be having more development impact than the commercial organisation that buses in parties of school children to re-paint classroom walls every six weeks. However, the private company may also have a significant impact through direct economic inputs through employment, payments to communities and purchase of local goods and services. It is also naive to discount the possibility that NGOs and charities can be inefficient and corrupt, just as private companies can be ethical and responsible (Bebbington *et al.*, 2008). Therefore it may be more beneficial to avoid this distinction on the basis of legal constitution and instead focus on the impact on the volunteer themselves and on the host community or organisation.

Impact on the volunteer

The primary impact at an individual level is on the volunteers themselves. They are the one having the experience of leaving the familiar and taking their goodwill and intentions to an unknown place. Jones categorises three main benefits to the volunteer: personal development, cross-cultural experience and global perspective (2005: 91). As such these may be good starting points for understanding the impact on the volunteer.

Personal development

The gap year has long been promoted on the basis of personal development with the argument that it allows students a time to mature, gain new experiences and develop new skills (Year Out Group, 2009). Increasingly universities are encouraging students to spend periods of time overseas developing their 'softer' skills. This fits with both an agenda for internationalisation and employability in Higher Education.

The Confederation of British Industry (CBI) identify 'soft skills' as being self-management, team working, problem solving, communication, application of literacy and business awareness (2009a: 13). Many of the respondents Jones identifies attest improvement in their soft skills to their short-term volunteer placement, as Jones summarises:

> Being in an unfamiliar linguistic and cultural environment forced cohorts to interact with each other more than they might have done in their home country or in a more culturally familiar setting, and many mentioned the

'greater stresses' placed on them by being in an overseas placement. In terms of confidence and interpersonal, problem-solving and communication skills, the transformative impact was more acute in the setting of a low-income country, which made the volunteering experience more challenging than it would have been in the home country.

(Jones, 2005: 93)

While a link between short-term volunteering and employability is difficult to prove, an improvement in the soft skills of young people is in line with the needs of employers. The CBI report that 78 per cent of employers look for soft skills when hiring graduates and 72 per cent look for positive attitude (CBI, 2009b: 24). As returned volunteers both identify their soft skills as having improved and are bound either for university or graduate employment, it seems sensible to suggest that there is meaningful impact to be had on the part of the volunteer. Measurement of impact on the volunteer in this regard could be achieved through self-evaluation by the volunteer pre- and post-placement.

Cross-cultural exchange and a global perspective

A significant potential impact on the individual can come through the opportunity for cross-cultural exchange that the volunteer experience can entail. The benefit of this process can to some extent be determined by the attitudes of volunteers themselves. They need to be patient and open to the new experiences, ready to learn and slow to judge. A very extensive literature on personal attitudes in development practitioners already exists and has been very influential in work to promote meaningful community participation in development. Influential authors in this area such as Robert Chambers talk of a new professionalism whereby the superior knowledge of the professional outsider is subservient to the local (Chambers, 2005). The impact of the experience on a volunteer and on their host can be improved if expectations and attitudes are managed more effectively. The neo-colonial discourse prevalent in much of the volunteer industry has not helped in this regard. The perception exists that volunteers with their Western knowledge and education can play a 'saviour' role. Certainly, an attitude of superiority is evident in some volunteers who believe that they have the answers, and do not necessarily understand the context in which they are operating (Erikson Baaz, 2005). This is not to argue that the volunteer cannot have a view on a situation, such as corporal punishment in schools, only that they critically reflect on the validity of that view.

Equally, contact with a resource-poor community in a developing country can give an individual volunteer a greater understanding of global inequalities and power, but also on different ways of living. It has become a well-research cliché for the returning volunteer to say 'well they were so poor but everyone was happy!' yet beneath the triteness of the expression are deeper questions of cultural and sustainable development. The volunteer of today is the community activist, CEO or government minister of tomorrow and they can shape the attitude of society to move beyond the neo-colonial myths. This is certainly one of the intentions behind the British government's recent support of volunteer tourism initiatives such as Platform2 – a programme aimed at disadvantaged young people in the UK to enable them to volunteer overseas (Platform2, 2009).

Therefore in returning to our question of how to enhance impact – in this regard preparation of the volunteer can emphasise attitudes and skills. It should also introduce them to the contextual background of the setting. In this regard good local support staff/volunteers can be crucial to helping a volunteer to manage their placement. Again this could be measured through self-evaluation or critical reflection by the volunteer. We use the term 'critical reflection' here in the sense indicated in Kolb's (1984) learning cycle as it requires experience to be related to theory. So when a volunteer asks themselves 'did I make a difference?', the answer they give is at least informed by some understanding. There is also potential for education providers to link volunteer tourism to assessment that promotes critical and reflective learning.

Impact on the host community

One of the more problematic aspects of talking about volunteer tourism is that most of the literature and criticism of volunteer tourism comes from research on the experience of the volunteer rather than on the host community (Mdee and Emmott, 2008). The crux of the neo-colonialist argument is that short-term volunteering or volunteer tourism has no benefits for the host community. VSO argue that some placements are 'badly planned' and 'spurious' (2007) but, assuming this is true (and without doubt it is true that some placements are badly planned and spurious), does that mean there is a negative impact on host communities? Or, is it simply that short-term volunteer tourism is somewhat less effective than its purveyors claim; or that it fails to lift communities out of poverty? We would like to assert that volunteer tourism or the efforts of individual volunteers are highly unlikely in themselves to 'lift' people out of poverty. That said, neither will the activities of VSO, although their experienced and highly prepared volunteers will make a contribution to the collective effort of development in any particular context.

We could certainly begin to understand the characteristics of volunteer tourism on host communities with the following categories: effective, ineffective and exploitative. These categorisations are developed from short periods of ethnographic research in Tanzania and Cambodia but need to be strengthened with further research. They are outlined here for the purpose of debate.

An *effective* volunteer scheme uses the money paid by volunteers to run their organisation in the host country and also to fund their/other development projects. They will have a long-term presence in those communities and will employ a majority of local staff (including in key decision-making positions). Volunteers may be matched with local counterparts (often students wanting to learn English). They will work on projects and in organisations where specific tasks have been defined for them. The organisation will manage volunteers' expectations and perceptions through training and support. In this way volunteer tourism can contribute to supporting local livelihoods (and many volunteers disregard this aspect). It is highly unlikely that any short-term volunteer will see for themselves large developmental benefits. Even if you build a classroom today you don't know if there will be sufficient teachers to staff it. The longer a volunteer stays the more likely they are to see and to understand the impacts of their engagement e.g. their class becomes more proficient in English, or the funding proposal they have been writing wins support. In effective organisations volunteers are prepared for interaction with the community, not necessarily that they are following prescriptive rules of behaviour but at least so they can greet and thank their hosts. Communities are themselves not homogenous and 'rules of behaviour' are also often a point of debate (Mdee, 2008). The process of cross-cultural exchange can enable such debates to happen in a community. Volunteers should not believe that communities exist in reified bubbles of 'tradition'. A more controversial question at this point might also raise the 'mission' of the organisation. A proselytising religious mission organisation may offer highly effective volunteer placements in relation to these criteria. A private company may also be able to contribute to effective development if it partners appropriately with local NGOs and enables them to access revenue for their projects and not just for hosting the volunteer.

Those that are *ineffective* may be so for a number of reasons; there may be inadequate communication between host and sender, poor planning of volunteer activities or a poor grasp of development issues by host, sender or both. In the case of the latter, we see a danger in volunteer operations being set up by the DIY developers, people with little experience of development work but with the laudable urge to 'do something', who can now easily and relatively cheaply (through cheap flights and email) begin operations. They may, for example, concentrate on building infrastructure without community participation and without employing locally or set-up orphanages without understanding the policy of the government is to support children within a family setting. They will be welcomed in by host communities who see potential for gaining access to resources, and potentially local elites may exploit the situation for their own benefit. This type of scheme may be run by profit or not-for-profit organisations. It should be noted also that many schemes can begin ineffectively but may become effective through a process of sustained engagement in that location.

Exploitative schemes may, and most likely will, fit into some or all of the above categories; however these schemes tend to charge the volunteer several thousand pounds for what is, in effect, a holiday, which offers to 'contribute towards a bright future' or 'build futures'. Exploitative schemes are not development focused; instead, they are motivated by profits, which can be readily made from what is a rapidly growing market.

Exploitative schemes are generally run by private companies on a profit-making basis and at present no mechanism exists through which to judge their claims of 'making a difference'. This is not to say, however, that exploitative schemes are either damaging host communities or morally illegitimate; they are exploitative because they use the promise of 'making a difference' and under-developed communities to extract profits. If the volunteer is aware of this, and the host community is not harmed (and presumably there will be some marginal economic gains e.g. $100 paid to a school to host a volunteer), then the market will dictate the demand for volunteer tourism placements. A question for further research is this: if all young people who undertake short-term volunteer placements in developing countries were aware of the crucial distinction between private and social enterprises in the market, how many would continue to choose such schemes?

As recruiting a paying volunteer is certainly a market transaction, then we accept that it can be a profit-making enterprise. In this sense there may be an argument that a social enterprise model might be an optimum arrangement for both balancing social and financial imperatives. The Cabinet Office defines social enterprise as 'businesses with primarily social objectives whose surpluses are principally reinvested for that purpose in the business or community, rather than being driven by the need to maximise profit for shareholders and owners' (DTI, 2002: 7). If a placement provider is driven by social aims, rather than private profit, then it would be paradoxical to suggest that they are exploitative. However, there is a great deal of debate in recent literature about the merits of social enterprise in both service delivery and as a means of enhancing community development and reducing poverty. While governments and donors are increasingly turning to social enterprise as a means of capturing the ability of markets to generate social development, its impact and ability to do this is highly uncertain (Toner *et al.*, 2008: 149). Social enterprise has, on the one hand, gained pragmatic and moral legitimacy among stakeholders, yet on the other empirical evidence of its enhancement of social outcomes is scant (Dart, 2004: 417–418).

Unfortunately, analysis of social impact by social enterprise is difficult to measure. We know they have to demonstrate 'added value', but no universally accepted method of measuring this has emerged. This chapter will not go into detailed discussion of these as this can be found elsewhere (see for example, Toner *et al.*, 2008; Nicholls, 2005); however, among favoured methods are social auditing, the balanced score card and social return on investment (SROI). While SROI is the most comprehensive measure, based on cost-effectiveness calculations, it is a complex tool to operate and beyond the capability of many social enterprise personnel, who would require specific training in using it effectively (Ryan and Lyne, 2008). For this chapter, and the recommendations it makes later, we will have to accept social auditing as an imperfect model of measuring social impact. Organisations such as Traidcraft and Furniture Resource Centre publish social accounts, which 'set out progress against specific strategic objectives such as number of people employed or those given additional healthcare or educational benefits' (Nicholls, 2005: 9). They are imperfect, however, because they are not able to quantify beyond a headline figure the value of social impact. That said, social auditing is a simple and relatively effective way for social enterprises to measure their own impact, without the help of external parties, or complex calculations (as with SROI), which, as we will show later on, of great practical value.

Towards an audit process for volunteer tourism

This chapter has, thus far, demonstrated that volunteer tourism is a rapidly growing, but inherently problematic market. It is relatively new, comparatively ill researched and completely unregulated. While we aim to contribute to the debate regarding short-term volunteer tourism, its primary objective is to outline a possible solution to many of the problems already detailed; one that avoids the neo-colonialist critique, prevents exploitation, both of host communities and volunteers, ensures a meaningful and sustainable impact, and, most of all, is development orientated. The ideas outlined below extrapolate from Mdee and Emmott's (2008) argument and start from the case of fair trade certification as an intellectual and practical basis for discussion.

Fair trade labelling

Fair trade labelling, certification and the 'movement' surrounding it are now well established. From its foundations in the late 1980s it has mushroomed, with the market for fair trade goods in the UK consisting of over 1500 products with a value of £290 million (Fairtrade, 2009). It has grown quickly and become so well established because retailers realised there was a demand for it; a niche market for ethical consumption which they could benefit from if they collaborated with one another. On the other hand, leading charitable organisations, such as Oxfam, World Development Movement and Christian Aid, were also involved in the early development of fair trade. This meant that, in its early stages, the fair trade movement was both consumer and producer driven and was as such able to engage stakeholders.

Fair trade has a relatively simple organisational structure. All fair trade products come under the umbrella of the fair trade labelling organisation (FLO). Beneath this, there are national initiatives, which are in charge of licensing producer groups to ensure they have a transparent and democratic structure, while at the same time negotiating a higher ('fairer') price for their products. The FLO's board is made up of six national initiative representatives, four producer representatives (one from Asia, two from Africa and one from Latin America), one alternative trade organisation representative and one business representative, ensuring representation for all stakeholders. Each group elects its representatives and the board meets quarterly. The FLO negotiates and agrees standards with national initiatives and then enforces them by inspecting producers and auditing supply chains. Standards enforced by the FLO come under three headings: producer organisational requirements, sustainable production requirements and trade standards. The rigorous standards applied by fair trade are all designed to guarantee the best deal for producers, even if this means that they have to change their operational practices in order to become certified (Nicholls and Opal, 2005).

The fair trade movement has certainly grown rapidly, but does this demonstrate a genuine positive step toward economic and social development in developing countries, or a 'fad' driven by well-meaning, but ineffective and ultimately aloof consumers in developed countries? Whilst there are no panaceas in development, there is some evidence at the micro level to suggest that fair trade has had a positive economic and social impact. In surveying the impact of fair trade in livelihood vulnerability in coffee producers in Nicaragua, Bacon (2005) shows that fair trade 'can help reduce livelihood vulnerability to the crisis in conventional coffee markets' (Nicholls and Opal, 2005: 508). Elsewhere, there are numerous examples of tentative support for fair trade, either at the conceptual level (see, for example, Reynolds, 2009) or on a micro level (Murray *et al.*, 2003, Nicholls and Opal, 2005).

Fair Trade Certification has generally been limited to food products. In relation to volunteer tourism perhaps the activities of the 'Year Out Group' for example show a process of 'certification' of gap year providers with their process of membership and limited external audit and scrutiny by an approval committee. However, their criteria are largely based on regulation of volunteer travel as a consumer experience for the volunteer and have no consideration of the impact on the host community.

Fair Trade is also now criticised as being too costly and restrictive with the benefits of certification being too marginal (Ellis and Keane, 2008). Certainly membership and certification of producers has kept many out of fair trade certification despite their potential development impact. This is because the producer bears the cost of compliance and high costs have kept small developing country producers out of fair-trade labelling schemes. This is also the case for small organisations in relation to membership of bodies such as the Year Out Group.

Good for development label

One recent response to the narrow, costly and exclusive nature of fair trade certification has been the suggested creation of a 'Good for Development' label (Ellis and Keane, 2008). Under this model the cost for producers is argued to be zero and is borne by the importers. The judgement of when a product is 'good' for development is determined by a broad view of development that assesses the value of increasing trade in specific commodities or products. If volunteer tourism were to adopt a 'good for development' label the potentially developmental aspects of the experience (ranging from the personal development of the volunteer to the provision of local employment) could be specified. Under the suggested good for development label, this would be a process of self-certification with different levels of accreditation (Ellis and Keane, 2008). In the case of volunteer tourism this would be a cost borne by the agent arranging the package. This may have knock-on effects for organisations based in developing countries who directly market themselves to volunteers but might still be of sufficiently low cost. It could also be used as a training tool with organisations to help them think through their development impact.

Self-certification: a responsible volunteering association (ReVA)

This section outlines a proposal for the creation of a Responsible Volunteering Association (ReVA) which was developed by a small UK volunteering NGO (Village-to-Village) based on a literature of current practice and discourse in the sector. The scheme is only at a conceptual stage and is used here to think through the possibilities. ReVA would in effect be a certification body that would establish 'standards' of practice in the area of voluntourism. Organisations could then apply to join ReVA on the basis of a self-audit against the standards which would build in progression (bronze, silver, gold). This is similar to the 'Good for Development' label outlined earlier. The aim is to educate both consumer and participants as to what might be 'good for development'. This would allow potential volunteers to compare 'general' schemes which are ostensibly geared towards poverty reduction, and more specific schemes, such as specific areas of conservation. It is proposed that ReVA be set up as a membership organisation, with fees associated upon joining and on a per volunteer basis. In return, joining members would become accredited. Once accredited, members would use a trademarked logo on their promotional information and in their marketing. Standards would be set and agreed upon relating to what information would need to be displayed on organisations' websites. There could also be different levels of accreditation, based on how well each placement provider meets the given criteria.

The aim of ReVA is to solve the problems of exploitative and ineffective schemes, discussed earlier, and to ensure economic and social benefits of short-term volunteering can be utilised in communities in developing countries. Ineffective schemes will be guided to improve their operations in order to meet the organisation's specifications. They will need, for example, to have open and demonstrable lines of communication and a long-term relationship with host organisations and communities; sound planning of each volunteer placement, impact assessments, matching and selective volunteer recruitment processes and an understanding of wider development issues. ReVA will only be open to development-oriented organisations and social enterprises. This is designed to actively promote development focused volunteer schemes and bring them to the forefront of potential volunteer consciousness. By promoting development focused volunteer schemes, it is hoped that ReVA will answer critics who have labelled volunteering as 'neo-colonialism'. This is because, rather than being a dominance of Western power and influence, as in the neo-colonial argument, placement providers will be those who are locally owned and managed, or who have genuine community links.

If planned and managed carefully, it is our belief that short placements of as little as four weeks can be of real benefit to host communities. This is in contrast to certain scholars and good quality volunteer scheme providers who argue that such short schemes have little effect on host communities. For example, an English teaching placement, which must carry on for a long period of time, would be suitable for a short-term volunteer as long as there is a syllabus in place and once one volunteer goes another one can replace them, continuing where they left. By doing this, assuming the programme is run by a local organisation, the community is placed at the centre of the scheme, yet the volunteer will still both benefit and make a positive impact.

Practical issues – beneficiaries

There have been other attempts at regulation, certification and good practice guidelines before in volunteer tourism (see, for example, Comhlamh, 2006) and to that end, ReVA does not pertain to be something new. However, none have gone to the lengths that this does and none have had the host community as their main beneficiary. Therefore, the primary target group to directly benefit from this proposal will be host communities. In order to become accredited, sending organisations and host organisations will have to be community focused. By being community focused, sending and host organisations will be able to identify the need of their communities and target volunteering to alleviate problems. In addition, the placement will be designed so the volunteer brings added value, rather than doing jobs which local people are capable of doing. This is possible in a number of areas, not least the English teaching example, mentioned above. Finally, by being locally 'owned', placements will facilitate a cultural exchange, rather than an imposition of Western values. While defending short-term volunteering against accusations of neo-colonialism, cultural exchange also helps the volunteer enhance their soft skills and cross-cultural awareness, an important benefit of volunteering, highlighted earlier.

The volunteer

The volunteer will benefit because they will be more able to make clear choices in their selection of a volunteer scheme. Organisations participating in ReVA should present full information on what impact the volunteer is expected to have. The volunteer will then be able to make a distinction between different types of scheme, something which, at present, is difficult to do. For example, if a volunteer is keen to work on conservation projects, sending organisations will be required to demonstrate both what conservation work the volunteer will be engaged in, what the intended impact of the scheme is, as well as more remedial information regarding what support is available in the country. They should also be able to see an example breakdown of how their money will be spent. All of this will allow volunteers to make more informed choices about which volunteer scheme suits them, and which is more in line with their desired impacts. Thanks to greater provision of information and project planning, the volunteer will have a greater knowledge of the development issues which face the region to which they are travelling. This will both broaden their knowledge and give them a more realistic picture of their contribution.

There is, at present, no shortage of choice for the volunteer when selecting a placement. However, it can be difficult to know what the day-to-day routine will be on placement and what its impact (both on an individual and community level) will be. ReVA presents an opportunity for placement providers to identify to potential volunteers what skills they will gain in terms of personal development, cross-cultural experience and global perspective, which, as identified earlier, are the main ways in which an individual develops. By giving volunteers this information, their expectations will also be managed more successfully. Evidence exists of volunteers being under-prepared and having over-ambitious expectations of what will be achieved during their placement (Mdee and Emmott, 2008). By labelling volunteer schemes, this expectation can be more successfully managed; volunteers can align their placement choice with their own personal development and can choose an impact area which suits their interests.

The sender

The sending and host organisations are also intended beneficiaries. First, by way of definition, the sending organisation (sender) is usually the UK (or other developed country) based organisation which recruits the volunteer. This is the volunteer's initial point of contact. The host organisation is the developing country organisation who the volunteer works for during their placement.

Much of the burden of ensuring that the many checks and criteria involved in joining ReVA is placed on the sender. This raises the important question of how the sender will benefit from this. First, once certified, the sender will be allowed to use the label, which shows that they are ReVA accredited. Use of this label will give the sender an advantage over non-accredited schemes because potential volunteers will associate the label with a good quality, development focused scheme which also protects their needs. Furthermore, senders will be keen to develop in this way, as if they are truly community focused, their development will inherently bring benefits for their target communities.

There will be clear lines of communication which allow the host to access suitable volunteers. They will also be able to make contact with the volunteer prior to the placement. This will allow for better preparation for all stakeholders, but will mean that the volunteer can come equipped. Like the sending organisations, hosts will need to be well organised, with clear decision-making structures and accountability in their operations. Once again, as the overall aim is to benefit the community, genuine organisations will see an alignment between their interests and those of the community.

ReVA, then, is a coming together of all stakeholders involved with short-term volunteering to ensure that the volunteer gets a fair deal and the developmental needs of the host community are, in some way, met. To reiterate, the aim cannot and should not be to promote short-term volunteering as a wide-ranging development solution, rather to see it as a beneficial activity. There are a number of ways this can occur. By ensuring local ownership, people and communities can use their local area as a means to secure the revenues associated with tourism. Rather than sustain a skewed distribution of volunteer providers in developed countries, ReVA attempts to facilitate local communities' use of their livelihood assets in the trend for volunteer tourism. Given the current size of the market, the predicted growth and employer gravitation towards soft skills, there is good reason to believe that volunteer tourism provides a sustainable enterprise opportunity for host communities, which again can be facilitated through ReVA. This argument can be seen as one which can re-orientate the market from its present 'neo-colonial' position, to one which ensures volunteer tourism is pro-poor, and representative of the people in host communities.

Conclusion

We have argued a current emphasis in the early literature on volunteer tourism has focussed on the consumer experience of the volunteer. This is highly problematic in that it tends to overlook the potential collective benefit of volunteer tourism for developing countries, given that it has can (when operated in the 'right' way) be highly pro-poor. In this chapter we deconstructed the benefits of volunteering to those for the volunteer and those for the host community. Our analysis of benefits in host communities is based on a background in research into community development rather than the tourism market but also direct experience in supporting volunteer tourism in Tanzania. However, the question of how volunteer tourism can benefit host communities does need further research.

We have argued that development-oriented volunteer tourism providers (be they NGOs or private enterprises) can bring benefits to the host community but we must not fool ourselves that the primary benefit is necessarily the labour of the volunteer. This is not to say either that an individual volunteer cannot make a difference, they can, but the 'difference' that they make has to be properly contextualised and understood by both host and volunteer. Instead, by focusing on the volunteer and the community as separate entities, we were able to see different impacts. The benefits to the volunteer are identified as personal development, cross-cultural experience and global perspective, all of which have the potential to build 'soft skills'. The possibility for overseas volunteering to build in processes of critical self-reflection would certainly make it useful as part of Higher Education programmes.

In a previous article, Mdee and Emmott (2008) explored the question of whether a fair trade label would make sense for improving the development impact of volunteer tourism. However, the cost and exclusive nature of such schemes did not seem appropriate. Therefore we drew on the recent idea of a 'good for development' label in order to explore how a Responsible Volunteering Association might effectively promote good choices by volunteers, manage their expectations and maximise benefits to the organisations and communities that host volunteers. This is very much a work in progress and requires support from high level stakeholders in order to be recognised and adopted.

Notes

1 'It' being Costa Rica.
2 Personal comment – ODI Associate

References

Bacon, C. (2005) Confronting the coffee crisis: can fair trade, organics, and speciality coffees reduce small scale farmer vulnerabilities in Northern Nicaragua. *World Development*, 33(3): 497–511.

Baldwin, N. (2007) Volunteer tourism: is it really helping? *Travel Weekly*, 1886: 14–15.

Bebbington, A., Hickey, S. and Mitlin, D. (2008) *Can NGOs Make a Difference? The Challenge of Development Alternatives*, London: Zed Books.

Brown, F. and Hall, D. (2008) Tourism and development in the global south: the issues, *Third World Quarterly*, 29: 5: 839–849.

CBI (2009a) *Shaping up for the Future: The Business Vision for Education and Skills*. London: CBI.

CBI (2009b) *Future Fit: Preparing Graduates for the World of Work*. London: CBI.

Chambers, R. (2005). *Ideas for Development*. London, Earthscan.

Comhlamh (2006) *Code of Good Practice for Sending Organisations: Principles and Indicators*. Dublin: Comlamh.

Dart, R. (2004) The legitimacy of social enterprise, *Non-profit Management and Leadership*, 14(4): 411–424.

DTI (Department of Trade and Industry) (2002) *Social Enterprise: A Strategy for Success*. London: DTI.

Ellis, K. and Keane, J. (2008) *A Review of Ethical Standards and Labels: Is There a Gap in the Market for a 'Good for Development' Label?* Working Paper No 297, London: Overseas Development Institute.

Erikson Baaz, M. (2005). *The Paternalism of Partnership*. London, Zed Books.

Fairtrade (2009) http://www.fairtrade.org.uk/what_is_fairtrade/history.aspx. Accessed 6 August 2009.

Goodwin, H. (2009) Reflections on 10 years of pro-poor tourism. *Journal of Policy Research in Tourism, Leisure and Events*, 1(1): 90–94.

i2i (2009) http://www.i-to-i.com/volunteer-work-with-children.html and http://www.i-to-i.com/kenya. Both accessed 3 August 2009.

Jones, A. (2005) Assessing international service programmes in two low income countries, *Voluntary Action*, 7(22): 87–99.

Kolb, D. (1984) *Experiential Learning*. Eaglewood Cliffs, NJ: Prentice Hall.

Kweka, J., Morrisey, O. and Blake, A. (2003) The economic potential of tourism in Tanzania, *Journal of International Development*, 15: 335–351.

McEwan, C. (2008) Post-colonialism. In V. Desai and R. Potter (eds) *The Companion to Development Studies*. London: Hodder.

Mdee, A. (2008) Towards a dynamic structure-agency framework: understanding individual participation in collective development activities in Uchira, Tanzania, *International Development and Planning Review*, 30(4): 399–421.

Mdee, A. and Emmott, R. (2008) Social enterprise with international impact: the case for fair trade certification of volunteer tourism, *Education, Knowledge and Economy* 2(3).

Murray, D., Reynolds, L. and Taylor, P. L. (2003) *One Cup at a Time: Poverty Alleviation and Fair Trade Coffee in Latin America*. Fair Trade Research Group, Colorado State University.

Nelson, F. (2004) *The Evolution and Impacts of Community-based Tourism in Northern Tanzania*, Issue Paper No. 131, London: Institute for Environment and Development.

Nicholls, A. (2005) *Measuring Impact in Social Entrepreneurship: New Accountabilities to Stakeholders and Investors*, Working Paper for the Skoll Foundation, Skoll Centre for Social Entrepreneurship, Oxford University. Available at http://www.sbs.ox.ac.uk/html/faculty_skoll_main.asp.

Nicholls, A. and Opal, C. (2005) *Fair Trade: Market Driven Ethical Consumption.* London: SAGE Publication Ltd.

Nkumah, K. (1965) *Neo-Colonialism: The Last Stage of Imperialism.* London. Nelson.

Platform2 (2009) http://www.myplatform2.com. Accessed 29 December 2009.

Power, S. (2007*) Gaps in Development: An Analysis of the UK International Volunteering Sector.* London: Tourism Concern.

Radcliffe, S. A. (2005) Re-thinking development. In P. Cloke, P. Crang and M. Goodwin (eds) (2nd edn) *Introducing Human Geographies.* London: Hodder.

Realgap (2009) https://www.realgap.co.uk/Ghana%20Gap%20year%20in%20Ghana. Accessed 3 August 2009.

Reynolds, L. T (2009) Fair Trade, *International Encyclopaedia of Human Geography.* Oxford: Elsevier, pp.8–13.

Ryan, P. and Lyne, I. (2008) Social enterprise and measuring social value: methodological issues with the calculation and application of the social return on investment, *Education, Knowledge and Economy*, 2(3): 223–237.

Simpson, K. (2004) Doing development: the gap year, volunteer tourist and a popular practice of development, *Journal of International Development*, 16: 681–692.

theleap (2009) http://www.theleap.co.uk/asia/cambodia-gap-year-team.html. Accessed 3 August 2009.

Toner, A., Lyne, I. and Ryan, P. (2008) Reaching the promised land: can social enterprise reduce social exclusion and empower communities, *Education Knowledge and Economy,* l2(1): 1–13.

VSO (2007) Gap year students told to forget aid projects. In *The Times*, 14 November.

Year Out Group (2009) http://www.yearoutgroup.org. Accessed 29 December 2009.

16 Volunteer tourism
Structuring the research agenda

Angela M. Benson

Introduction

The early literature on volunteer tourism tended to be exploratory in that it tried to demonstrate understanding and knowledge of a new and complex phenomenon. The more recent literature has attempted to go beyond this and develop a more structured and theoretical approach. While theoretical frameworks of volunteer tourism are still in their infancy, both in terms of adopting existing theoretical frameworks and developing new ones, it is clear that the growing body of researchers (chapter authors and other academics) in this area are seeking a more critical discourse associated with volunteer tourism. Practical application is evident in many of the chapters of this book (and in related journal articles). Whilst this aids in the building of theory and critical discourses, it is unclear the extent to which knowledge transfer has taken place in a wider context, but evident that without the cooperation of individuals and organisations much of this empirical work would have been impossible to undertake. In order to build on existing results and address the gaps in the existing literature, this chapter continues by offering a structure for a future research agenda.

Structuring the research agenda is divided into three sections. The first section examines volunteer tourism as part of the bigger picture and suggests five areas. The second section, critical issues in the volunteer tourism sector, introduces six areas and finally, the third section outlines four key stakeholders.

Volunteer tourism as part of the bigger picture

Linking to other disciplines

The discussions surrounding cross-disciplinary and multi-disciplinary working are common themes in academic circles. However, the extent to which tourism and more specifically volunteer tourism is embraced by other disciplines and vice versa is questionable. There are some instances where this is happening. For example, Chapters 14 and 15 in this book draw from literature on development in order to understand the extent to which volunteer tourism makes a difference.

However, when examining the voluntary agencies and associated literature on volunteering in the UK, it is evident that the concept of volunteer tourism is barely acknowledged despite the prediction of growth both in size of value (see Chapter 1). For example, a recent book entitled *Volunteering and Society in the 21st Century* (Rochester *et al.*, 2010) mentions volunteer tourism once (p.111), although there is rather more discussion (pp.13–15) on serious leisure (Stebbins 1992; Stebbins 1996; Stebbins and Graham 2004; Stebbins 2007), even if the links between the two are not articulated. In order for the discussions on volunteer tourism to be theoretically robust it is important that any future research agenda must include being linked to and underpinned by other disciplines.

The third sector

The voluntary or not-for-profit sector is increasingly being referred to as the *third sector* and incorporates voluntary and community organisations, social enterprises, charities, cooperatives and mutuals. It has a voice in almost every country of the world and is seen to play a significant role in development. It is purported that the third sector has seen substantial growth in the last decade, with higher levels of professionalisation and with millions of people volunteering annually. It contributes to improving and developing communities and society, adds to the economy through thousands of organisations, operates in a sustainable framework and, consequently, there is an increasing acknowledgement of its worth. However, it is unclear the extent to which this rhetoric supports or conflicts with volunteer tourism, in that many of the organisations offering the volunteer tourism product would be classified under the third sector label. Whilst there is little literature on individual organisations, there is even less on volunteer organisations working collectively to improve and develop at a countrywide level. For example, there are a number of charities and non-governmental organisations based in the UK and active in Madagascar. A number of these are volunteer organisations (Azafady; Blue Ventures, Dodwell Trust; Earthwatch; Frontier). To what extent do the volunteer organisations (a) engage with other third sector organisations operating in Madagascar and (b) engage with each other?

Sustainability

Whilst volunteer tourism is often associated with the concept of alternative tourism, the discussion of volunteer tourism within the sustainable tourism development paradigm is limited and problematic. Volunteer tourism is only associated with or measured against the three fundamental pillars of sustainability – economic, social and environmental – to a limited extent. This is true within both the volunteer tourism literature and the wider tourism literature. Most case examples have been undertaken by focusing on a single dimension (or impact) at one moment in time. This is due in part to different dimensions of impact evaluation requiring their own methodologies and capabilities. Consequently, tools, methodologies and capabilities for taking a holistic approach to evaluating the combined social, economic and environmental contributions are in their infancy. Despite this the literature clearly indicates that balance between the three pillars is essential if sustainability is to be achieved (Cronin, 1990; Burns and Holden, 1995; Clarke, 1997; Swarbrooke, 1999; Hardy *et al.*, 2002; Ko, 2005; United Nationals Development Programme, 2005). Any impact research conducted on just one element is insufficiently rich to create meaningful data upon which to make informed management decisions.

Climate change

The tourism and travel industry are coming under scrutiny due to the environmental damage caused by air travel. Potentially, this issue could become a dichotomy for the volunteer tourism sector in that organisations promote themselves as part of a sustainable agenda but take volunteers thousands of miles from home for relatively limited periods and using air travel that contributes to climate change. Whilst many of the organisations themselves are not directly involved with volunteer travel arrangements and volunteers meet the organisations in the final destination country, nevertheless companies are complicit in the need to travel to projects. At a recent conference held by a research volunteer organisation for volunteers (past, present and future) the following comment, by an older female volunteer, was overheard 'I will only do volunteer projects in Europe now, where I can catch a train to. That's my personal contribution to helping with this global warming problem.' What is unclear is the extent to which other volunteers mirror this perspective. One volunteer organisation has commenced a carbon offsetting scheme: 'Blue Ventures Carbon Offsetting policy requires all volunteers, researchers, tourists and recreational visitors to our research site to offset the costs of national and international flights that have been used in getting to Andavadoaka' (Blue Ventures, 2007). How common is this within volunteer tourism? Are schemes such as this a result of organisations being proactive? Or reactive and therefore, due to volunteers pushing this agenda forward? Does this type of proactively by organisations offer competitive advantage in the marketplace?

Volunteering and other tourism niches

Volunteer tourism is increasingly becoming associated with other tourism niches (e.g. backpacking (Ooi and Laing, 2010)). The niche of sport tourism has been chosen here to be examined in detail as an example of structuring ideas to analysis other niches that are often linked to, or overlap with, volunteer tourism. The study of volunteering and sport is well documented within the Western world whilst the linking of volunteering and sport tourism is more recent with much of the research focusing around events, in particular mega events (Olympics, World Cups). However, according to Baum and Lockstone (2007) even this area lacks a holistic approach. The field of volunteering and sport tourism demonstrates that a range of opportunities and challenges are emerging. For instance, a growing number of volunteer tourism organisations are offering 'sport volunteer projects overseas'; colleges and universities are travelling with volunteer sport students to engage with communities in a sporting context; sport tourism volunteering is occurring at events, with volunteers travelling both domestically and overseas to take part. These burgeoning opportunities however, raise a plethora of questions and issues and it is evident that the current literature offers few answers. In short, little is known about volunteering in a sport tourism context. Consequently, the areas to research are extensive, with the list below being just an indication:

- Understanding the volunteer in sport tourism (who is the volunteer in regards to their behaviour, motivation, experience, gender, contribution, impact?). Intercultural perspectives on sport tourism and volunteering (one advert states that 'sport is a universal language'; is this true? If so, what effect does it have on adaptation, culture confusion and cultural exchange? If not, what engagement is happening?
- Supply side – which sectors are involved: private, public or third sector organisations? To what extent are partnerships being formed? Is this just an extension of existing companies portfolios of projects or is it an opportunity for new entrants to enter the market?
- Sponsorship, funding and payment – how is volunteering in sport tourism being funded?
- Legacy of volunteering in sport tourism – tangible and intangible – whose legacy: the country where the volunteering took place or the country the volunteers return to? To what extent do relationships continue after volunteers return home? What is the extent of serial volunteering in sports tourism?

Critical issues in the volunteer tourism sector

Definitions, boundaries, terminology and measurement

There is no real agreement over a definition of volunteer tourism. This being said, the most commonly cited definition of volunteer tourism is that of Wearing (2001), which is reinforced by its usage in this book. However, Lyons and Wearing (2008) acknowledge that this definition is now limiting and other definitions have begun to emerge, but none appears to capture the myriad of variables that need to be encompassed. In addition, the boundaries of volunteer tourism are often blurred as the debates of 'what is' and 'what is not' volunteer tourism continues. The issue of terminology is also pertinent here. For example are we defining volunteer tourism or voluntourism, or both? It is not the intention to offer a definition here but to suggest ideas that could be considered for inclusion in the debate:

- Many of the definitions on offer are conceptual definitions of volunteer tourism which attempt to provide a theoretical framework by which to identify its essential characteristics. However, conceptual reasons are just part of the definition debate.

- From the practical perspective, one of the key reasons for seeking a definition is in order to achieve meaningful statistical data that can be compared world-wide. Technical definitions would provide tourism information for statistical or legislative purposes and in order to achieve this, the definitions tend to include a time and/or distance element.
- Whose 'voice' is used and therefore, the consideration of whether definitions are orientated to the supply-side or the demand-side
- Definitions of both volunteering and tourism should be examined, if only to discard them at a later date! For example: a key determinant when examining or defining the act of 'volunteering' is that within the definitions, the emphasis is on nominal payment or expenses to the volunteer, as indicated in the definition by Cnaan *et al.* (1996) who used the term *remuneration*. In the definition by Stebbins and Graham (2004: 5) 'no or, at most token pay' is used and in the UN definition the terms *reimbursement of expenses and some token payment* (UN, 2001). None of these encompass the issue of payment by the volunteer which is often associated with the concept of volunteer tourism.
- Early definitions of volunteer tourism have tended to favour the altruistic side of volunteering. More recent debates, however, ask 'Is it (volunteer tourism) altruistic or ego-centric?' (or one of many positions along that continuum). A more common view is now being recognised that most forms of volunteer tourism (and indeed, modern volunteering) are not just about 'doing good for others' but also about 'doing good for self' (Matthews, 2008: 111) which is echoed in the UN definition of volunteering (UN, 2001).

The question of money

Whilst it is now generally accepted in the literature that volunteer tourism is often associated with a 'payment' by the volunteer, the discussions are perfunctory and fleeting. There is little empirical research related to the issue of volunteers paying. For example, how does the concept of volunteers paying fit in to mainstream volunteering theoretical frameworks? Where does this money go? Is it clear and transparent of how the money is spent? Is the project value for money and from whose perspective – for the project; for the volunteer; for the supplier of volunteer opportunities; for the host community? What proportion goes to the host community? Do volunteer expectations change as a result of paying? Do they see themselves as a customer foremost and a volunteer second or vice versa? How do volunteers fund their expeditions? This then leads to the issue of fundraising; whilst it is barely acknowledged in the literature, it is very evident from organisational material and websites that fundraising is an option available to volunteers. Organisations offer practical guidelines and support mechanisms for volunteers to engage in fundraising activities either individually or as groups (e.g. from the same university). Whilst fundraising per se is not an issue, and is common place when linked to charitable organisations, its role and functions in volunteer tourism are yet to be explored.

There are also a series of questions around volunteers as spending tourists and to what extent this contributes to the economies at local (host community), regional and country levels. Do volunteers extend their travel arrangements prior to and/or after the volunteer activity? If so, is this activity at the luxury end of the market or backpacking? How long do they travel for? And therefore, could the volunteer contribution be significant or is it minimal?

Volunteer tourism as a quality product

Whilst there are lists about the varied destinations, the types of programmes and even clustering of programmes/projects, it is only recently that the discussions have taken a more critical viewpoint and questioned the extent to which the projects are 'worthwhile' and 'if they make a difference' and to whom, the volunteer, the host community? Questions such as 'Are organisations interested in the extent to which their projects are "worthwhile"?' are yet to be addressed. Engaging in quality can include both internal and external measures. However, the discussions at either organisational level or sector level are lacking. There are a number of tools that are used in mainstream tourism for measuring or monitoring and evaluating – SERVQUAL, performance indicators, benchmarking, certification and accreditation (Chapter 15 for example outlines work that is in progress associated with accreditation) – but generally, their usage in volunteer tourism is limited. However, critical discourse in this area is crucial if volunteer tourism is to be seen as an ethical form of travelling.

Corporate responsibility

Corporate responsibility has strong arguments for it and against it. Regardless of on which side of the fence you sit, it is high on the management agenda. While much of the corporate responsibility falls under mainstream volunteering and philanthropy, there are volunteer tourism projects used by corporations which enable their employees to engage in a worthwhile project. The volunteer payment is made by the corporation to the organisation on behalf of the employee, albeit at a discounted price depending on the negotiations that have taken place. For example, the Earthwatch Institute uses volunteers to collect field data for scientists for a range of projects. Earthwatch Europe have been particularly successful at developing corporate partnerships, one of which is with the HSBC Climate Partnership. The partnership was for a period of five years and consisted of HSBC employees (in total 2,200 employees) from across the business spending two weeks at one of the five Regional Climate Centres that were set up around the globe to carry out field work to establish the health of the forests. This appears to be a 'win-win' situation, as the companies have an opportunity to demonstrate their corporate social responsibility and Earthwatch furthers its aims and objectives. In 2006/07, 70 per cent of the Earthwatch USA office income was generated from volunteer funds whereas for the Earthwatch Europe office, volunteer funds contributed only 20 per cent, with the rest coming from corporate partnerships, foundations, trusts and other accessible charity grants. This potentially has far-reaching implications. This part of the market is growing, but little is known about corporate responsibility and its role with volunteer tourism.

Technology

The development of mobile phones, the internet, blogs and other major global social networks (Facebook, Twitter, YouTube and MySpace) have made travelling easier through this connected world. O'Regan (2008: 113) argues that 'the world today is so networked that strangers no longer exist, but are simply connections waiting to happen'. Whilst the literature relating to tourism and technology is apparent (Mascheroni, 2007; White and White, 2007), currently there is no reference to technology and its use by volunteers in the volunteer tourism literature. Is technology also furthering the volunteer tourism agenda?

Risk

Recently, Benson and Siebert (2009) identified that crime, violence and poverty were associated with 'worst' experiences by volunteers. Consequently, 'It becomes increasingly clear that the "risks to" and "safety of" volunteers is becoming a more prominent issue; it is evident from this study, a few other studies, and reports in the press, that volunteering is not without its dangers' (Benson and Siebert, 2009: 310). In this study a number of questions were raised, which still need addressing and therefore, are reiterated here:

- To what extent do the 'worst' experiences influence future volunteer potential?
- Is the image of the destination affected and are return trips to a destination and 'word-of-mouth' marketing to friends and family affected or do the volunteers see it as part of the experience?
- This then raises questions about the extent to which organisations make volunteers aware of potential dangers and what mechanisms are put in place to address these issues or is it up to the volunteer to become 'streetwise' in the chosen destination?

Chapter 12 of this book has also identified risk as an important issue and offers a way forward in the risk management process for organisations.

Volunteer tourism and key stakeholders

Volunteers

While the focus of the research has tended to be on the volunteer, there are still areas that have little or limited research, two of which are discussed here.

Segmentation studies comprise a large part of the tourism literature due to the size and complexity of the travel and tourism sector. There is an extensive range of typologies of tourists from 1970 onwards; the most often cited typologies are Gray (1970), Cohen (1972), Plog (1974), Cohen (1979), Smith (1989) and Plog (1994). As the move towards alternative tourism products became evident, studies have been undertaken to examine the typologies of more environmentally aware tourists (Lindberg, 1991; McCool and Reilly, 1993; Palacio and McCool, 1997). Despite this richness in tourism literature, the literature on typologies contextualised within volunteer tourism is minimal. Clifton and Benson (2006) profiled a group of Research Volunteer Ecotourists in Indonesia by motivational statements; the segments included culturally orientated; experientially orientated; socially orientated; personal achievers and relaxation orientated, and more recently, McGehee *et al.* (2009) segmented volunteer tourists into three clusters: vanguards, pragmatists and questers. As indicated in the first chapter, the volunteer tourism market place has already become segmented by market providers; associated research, however, is lagging behind. Dann *et al.* (1988: 10) believed that typologies have contributed to moving tourism research to higher levels of theoretical awareness. It is anticipated that research using a life-style typologies approach will act as a similar catalyst for volunteer tourism.

This book has two chapters (7 and 8) which examine volunteers retrospectively; another chapter (3) discuss the issue of longevity of volunteering projects. However, a longitudinal research gaze within the volunteer tourism context is missing. There is criticism about the short-term placements of volunteers. However, there is little discussion to balance these viewpoints regarding long-term projects. Some companies have been taking volunteers annually to a destination for ten years or more (e.g. Operation Wallacea has been operating in Indonesia since 1995); but little is known about the accumulative affect, positive or negative, on the various stakeholders. With the growth of volunteer tourism and new companies entering the marketplace, there is an opportunity to engage in longitudinal studies as, by extending the duration of research engagement, the potential of greater understanding involved in social change becomes a viable discourse.

Host communities

Whilst there is extensive literature relating to host communities within the broader tourism literature as highlighted in Chapter 1, the discussion of volunteer tourism and host communities is negligible and, consequently, the question of what should be researched is open-ended.

Governments

All of which might suggest that Malawi is off the beaten track. Wrong. The place is swarming with visitors, and almost every single one is with an *organization*. They are volunteer tourists – or … 'voluntourists' – and … I was one of them.

(MacKinnon, 2009)

While it is clear that volunteer tourism is a worldwide phenomenon, it is unclear the extent to which governments are involved. Are they passive recipients of pro-gammes in their countries or are they actively engaged in using volunteer tourism as a vehicle for capacity building, poverty alleviation, peace etc.? Anecdotally, the answer is both, however, empirical evidence is lacking. The TRAM report (Tourism Research and Marketing (TRAM), 2008) gives some examples of gov-ernment and tourist board involvement, but generally the area is under-researched and leaves many questions to be answered. For example: which governments include volunteer tourism as a niche as part of their tourism development strategy? Is there any coordination of a number of volunteer organisations in one country? Is there any attempt by government departments to get organisations together to enhance their own strategies? To what extent do organisations disseminate results (for example conservation and archaeology projects) in the country of data collec-tion? On what premise do volunteers enter a country (this is particularly pertinent when visas are involved)? For example, when applying for a visa to enter Ghana as a UK citizen there is now a box for volunteers to tick; to what extent have other countries adopted this approach?

Organisations

The literature related to organisational management and behaviour is extensive, volunteer programme management as a professional field is over 40 years old. However, the extent to which these areas have been applied to the volunteer tour-ism sector is minimal and consequently, as with the host communities (see earlier) the agenda for research on the organisations involved in volunteer tourism is open ended. This being said, there are three areas, to which particular attention has been drawn.

- *Type of businesses*: the volunteer tourism sector has both private and third sector organisations. Do private volunteer tourism companies differ from not-for profit organisations, or those run as a social enterprise? Do volunteers choose their project on the basis of the type of business? Charitable organisa-tions have access to funding that private organisations do not. To what extent have volunteer tourism organisations accessed this funding and what impli-cations has this had at country/regional level?
- *Management*: while this topic is immense – strategic intent, leadership, shareholders, staffing, etc. – it is necessary to point out that this area is virtu-ally untouched in respect of volunteer tourism. While general questions on management abound, there are numerous areas where the particular charac-teristics of management in an organisation relating to the third sector need reinterpreting in the context of volunteer tourism. For example, taking staff development as an individual component of management, some more spe-cific questions would include: 'Do volunteers go on to become managers of volunteer tourism projects or organisations?'; and 'Is there a career in vol-unteer tourism as there is in other forms of third sector activities?'. Similar

mappings of general management issues to the specific context will yield a rich vein for additional research in volunteer tourism.

- *Professional behaviour and expectations*: there is no doubt that there are expectations from both volunteers and organisations, but to what extent are these in harmony? Volunteer codes of conduct or similar are not unusual, but do volunteers adhere to them? Is the process formalised? Do contracts exist between the volunteer and the organisation? One of the ways this has been explored is by examining the psychological contract between volunteers and organisations (Farmer and Fedor, 1999; Liao-Troth, 2001; Thompson and Bunderson, 2003; Blackman and Benson, 2010); however, even this is limited and the sector would benefit from a wider range of theories being used to examine the questions outlined.

Conclusion

Despite the growing popularity of volunteer tourism, systematic academic research in this area is still in its infancy. There is no doubt that the concept of volunteer tourism and our understanding of it have developed over a relatively short period of time but there is still lots of work to be done in order to fully appreciate and understand the value of this dynamic sector. It is important that we (the research community) encourage theory and practice to become entwined in that practice helps to build theory and theory improves, changes, impacts on or influences practice. This symbiotic relationship then between theory and practice would provide an effective framework which to promote successful organisations, deliver projects that accrue benefit for local communities and recognise the role of the volunteer. The understanding of these issues, gained through targeted research and disseminated to stakeholder groups, should ensure that managers are properly informed and equipped for effective management.

This chapter has offered an approach to structuring a research agenda. However, there is no pretence or expectation that the agenda described here is comprehensive or complete. There will be, without doubt, numerous detailed omissions with respect to the wide-ranging issues associated with volunteer tourism. Its purpose was not to be wholly inclusive but to offer a framework for ideas, and to generate discussion in the hope that other academics and professionals will engage in the process by adding to this structure and creating their own avenues for investigation.

References

Benson, A. M. and Siebert, N. (2009) Volunteer tourism: motivations of German participants in South Africa. *Annals of Leisure Research*, 23(3&4): 295–314.

Blackman, D. and Benson, A. M. (2010) Research volunteer tourism: the role of the psychological contract in managing research volunteer tourism. *Journal of Travel and Tourism Marketing*, 27(3): 1–15.

Blue Ventures (2007) http://www.blueventures.org.

Baum, T. and Lockstone, L. (2007) Volunteers and mega sporting events: developing a research framework. *International Journal of Event Management Research*, 3(1): 29–41.

Burns, P. M. and Holden, A (1995) *Tourism: A New Perspective*. Hitchin: Prentice-Hall.

Clarke, J. (1997) A framework of approaches to sustainable tourism. *Journal of Sustainable Tourism*, 5(3): 224–233.

Clifton, J. and Benson, A. M. (2006) Planning for sustainable ecotourism: the case of research ecotourism in developing country destinations. *Journal of Sustainable Tourism*, 14(3): 238–254.

Cnaan, R. A., Handy, F. and Wadsworth, M. (1996) Defining who is a volunteer: conceptual and empirical considerations. *Nonprofit and Voluntary Sector Quarterly*, 25: 364–383.

Cohen, E. (1972) Towards a sociology of international tourism. *Social Research*, 39(1): 164–182.

Cohen, E. (1979) A phenomenology of tourist experiences. *Sociology*, 13: 179–201.

Cronin, L. (1990) A strategy for tourism and sustainable developments. *World Leisure and Recreation*, 32(3): 12–18.

Dann, G. M., Nash, S. D. and Pearce, D. (1988) Methodology in tourism research. *Annals of Tourism Research*, 15(1): 1–28.

Farmer, S. M. and Fedor, D. B. (1999) Volunteer participation and withdrawal: a psychological contract perspective on the role of expectations and organisational support. *Nonprofit Management & Leadership*, 9(4): 349–367.

Ghobadian, A. (2010) Growing gulf between managers and research. *Financial Times*, 31 May 2010: 13.

Gray, H. (1970) *International Travel-International Trade*. Lexington DC: Heath.

Hardy, A., Beeton, R. J. S. and Pearson, L. (2002) Sustainable tourism: an overview of the concept and its position in relation to conceptualisations of tourism. *Journal of Sustainable Tourism*, 10(6): 475–496.

Ko, T. G. (2005) Development of a tourism sustainability assessment procedure: a conceptual approach. *Tourism Management*, 26: 431–445.

Liao-Troth, M. A. (2001) Attitude differences between paid workers and volunteers. *Nonprofit Management & Leadership*, 11(4): 423–442.

Lindberg, K. (1991) *Policies for Maximising Nature Tourism's Ecological and Economic Benefits*. Washington, DC: World Resources Institute.

Lyons, K. D. and Wearing, S. (2008) *Volunteer Tourism as Alternative Tourism: Journeys Beyond Otherness*. In K. D. Lyons and S. Wearing, *Journeys of Discovery in Volunteer Tourism*. Wallingford, UK: CAB International.

MacKinnon, J. B. (2009) *The Dark Side of Volunteer Tourism*, UTNE Reader. Available from http://www.utne.com/Politics/The-Dark-Side-of-Volunteer-Tourism-Voluntourism.aspx. Accessed 26 February 2010.

Mascheroni, G. (2007) 'Global nomads" network and mobile sociality: exploring new media uses on the move. *Information, Communication and Society*, 10(4): 527–546.

Matthews, A. (2008) Negotiated selves: exploring the impact of local-global interactions on young volunteer travellers. In K. D. Lyons and S. Wearing, *Journeys of Discover in Volunteer Tourism*. Wallingford, UK: CAB International, pp. 101–117.

McCool, S. F. and Reilly, M (1993) Benefit segmentation analysis of state park visitor setting preferences and behaviour. *Journal of Park and Recreation Administration*, 11(4): 1–14.

McGehee, N. G., Clemmons, D. and Lee, S. J. (2009) 2008 Voluntourism Survey. Available from http://www.blurb.com/bookstore/detail/878048. Accessed 14 February 2010.

Ooi, N. and Laing, J. H. (2010) Backpacker tourism: sustainable and purposeful? Investigating the overlap between backpacker tourism and volunteer tourism motivations. *Journal of Sustainable Tourism*, 18(2): 191–206.

O'Regan, M. (2008) Hypermobility in backpacker lifestyles: the emergence of the internet café. In P. Burns and M. Novelli, *Tourism & Mobilites: Local-Global Connections*. Wallingford, UK: CAB International.

Palacio, V. and McCool, S. F. (1997) Identifying ecotourists in Belize through benefit segmentation: a preliminary analysis. *Journal of Sustainable Tourism*, 5(3): 234–243.

Plog, S. C. (1974) Why destination areas rise and fall in popularity. *cornell hotel and restaurant administrative quarterly*, 14(4): 55–58.

Plog, S. C. (1994) Developing and Using Psychographics in Tourism Research. In J. R. Brent Ritchie and C. R. Goeldner, *Travel, Tourism, and Hospitality Research*. New York: John Wiley & Sons, Inc: 209–218.

Rochester, C., Ellis Paine, A., Howlett, S. and with Zimmeck, M. (2010) *Volunteering and Society in the 21st Century*. Basingstoke, UK: Palgrave MacMillan.

Smith, S. L. J. (1989) *Tourism Analysis*. Harlow: Longman.

Stebbins, R. A. (1992) *Amateurs Professionals and Serious Leisure*. Ulster: McGill – Queens University Press.

Stebbins, R. A. (1996) Volunteering: a serious leisure perspective. *Nonprofit and Voluntary Sector Quarterly,* 25(2): 211–224.

Stebbins, R. A. (2007) *Serious Leisure: A Perspective for Our Time*. New Brunswick, New Jersey: Transaction Publishers.

Stebbins, R. A. and Graham, M. M. (2004) *Volunteering as Leisure/Leisure as Volunteering.* Wallingford, Oxon, UK: CAB International.

Swarbrooke, J. (1999) *Sustainable Tourism Management*. Oxford: CABI Publishing.

Thompson, J. A. and Bunderson, J. S (2003) Violations of principle: ideological currency in the psychological contract. *Academy of Management Review*, 28(4): 571–586.

Tourism Research and Marketing (TRAM) (2008) *Volunteer Tourism: A Global Analysis*. Barcelona: ATLAS Publications.

UN (2001) *United Nations Volunteers Report,* prepared for the UN General Assembly Special Session on Social Development, Geneva, February 2001.

United Nationals Development Programme (2005) T*he Sustainable Difference: Energy and Environment to Achieve the MDGs*. New York, United National Development Programme, Energy Environment Bureau for Development Policy, Energy and Environment Group: 1–346.

Wearing, S. (2001) *Volunteer Tourism: Experiences that Make a Difference.* Wallingford: CABI Publishing.

White, N. and P. White (2007) Home & away: tourists in a connected world. *Annals of Tourism Research*, 34(1): 88–104.

Index